THE 2011 NATIONAL ELECTRICAL CODE BOOK OF IN-DEPTH CALCULATIONS

VOLUME 2

*Covering selected Articles
of Chapters 1 through 3*

Alvin J. Walker

THE 2011 NATIONAL ELECTRICAL CODE BOOK
OF IN-DEPTH CALCULATIONS

VOLUME 2

© 2015 Alvin J. Walker

ISBN 13: 978-0-9831358-3-8

LCCN 2015907070
First Edition
1 2 3 4 5 6 7 8 9 10

Walker & Walker Electrical Consultants
For more information, please contact
Alvin Walker
318-393-6841
www.alvinwalker.com

TABEL OF CONTENTS

INTRODUCTION TO VOLUME 2 . . . . **xi**

Number in brackets indicates the number of questions per NEC sections. Total questions: 254

CHAPTER 1 - General

ARTICLE 110
Requirements for Electrical Installations

110.14(C)(1) 1
110.14(C)(1)(a) [3]. 1
110.14(C)(1)(b) [3] 3

CHAPTER 2 - Wiring and Protection

ARTICLE 210
Branch Circuits

210.3 5

210.11(A) 6

210.19(A)(1) [5] 6
210.19(A)(1) Informational Note No. 4 9
210.10(A)(3) 9
210.19(A)(3), *Exception No. 1*. 9
210.10(A)(3), *Exception No. 2*. 9

210.20(A) 9

210.23 10

210.23(A)(1) 10

210.23(B) [2]. 11

Table 210.24 12

ARTICLE 215
Feeders

215.2(A) Informational Note No. 2 13
215.2(A)(1) [4]. 13

Services [NEC References and Requirements Illustrated] .16

ARTICLE 230
Services

230.42(A) [2] 17
230.42(C) 18

230.79(C) 18

230.90(A) [2] 18

Overcurrent Protection References per Articles 20

ARTICLE 240
Overcurrent Protection

240.3 21

240.4(B) [6] 21

240.4(C) [3] 25

240.4(D) 26

240.4(E) 27
240.4(E)(1). 27
240.4(E)(4). 27
240.4(E)(6). 27

240.4(G) 27

Tap Conductors 28

240.21(B)(1) [3] 31
240.21(B)(2) [2] 34
240.21(B)(3) 35
240.21(B)(4) 38
240.21(B)(5) 40

240.21(C)(1) 44
240.21(C)(2) 45
240.21(C)(3) 46
240.21(C)(4) 47
240.21(C)(5) 47
240.21(C)(6) [3] 47

240.91(B) 51
240.92(B) [3]. 52

ARTICLE 250
Grounding and Bonding

250.24(C)(1) [6] 58

250.24(C)(2) [2] 60

250.28(D) [5] 60

250.52(A) 63

Table 250.64(D)(1) 63

Table 250.66 [7] 64

250.102(C)(1) [2] 66
250.102(C)(2) [3] 67
250.102(C)(3) 69

250.102(D) [4] 69

250.122(A) [2] 72

250.122(B) 72

250.122(C) 73

250.122(D) 73

250.122(F) 74

250.122(G) 74

CHAPTER 3 - Wiring Methods and Materials

ARTICLE 300
Wiring Methods

300.34 75

Ampacity Tables of Article 310 76

ARTICLE 310
Conductors for General Wiring

Parallel Conductors 77
Sizing Parallel Conductors. 77
310.10(H)(1) [2] 79
310.15(A)(2) 81

310.15(A)(2), *Exception* 82

310.15(A)(3) 83

Table 310.15(B)(16) 84
    What Determines A Conductor's Ampacity?
    Understanding Table 310.15(B)(16) formerly Table 310.16 85
    Understanding Table 310.15(B)(2)(A) formerly Correction Factors to Table 310.16. 87
    Problem-Solving Methods Based on Operating Conditions. 88
      Table 310.15(B)(16)-**METHOD 1** .89
      Table 310.15(B)(2)(a)/NEC 310.15(B)(3)(a)-**METHOD 2** 91
      Table 310.15(B)(16)/NEC and Table 310.15(B)(3)(a)-**METHOD 3** 93
      Table 310.15(B)(16)/NEC and Table 310.15(B)(3)(a)-**METHOD 4** 95
        Examples Based on The Conditions Of **METHOD 1** [2] 96
        Examples Based on The Conditions Of **METHOD 2** [2] 97
        Examples Based on The Conditions Of **METHOD 3** [2] 98
        Examples Based on The Conditions Of **METHOD 4** [6] 99

310.15(B)(2) 103

310.15(B)(3)(a) 105
310.15(B)(3)(a) Informational Note No. 1 [2] 106
310.15(B)(3)(a)(1) 107
310.15(B)(3)(a)(2) 107
310.15(B)(3)(a)(3) 107
310.15(B)(3)(c) 107

310.15(B)(4) 108

The Neutral Conductor 109
  310.15(B)(5) 109
  310.15(B)(5)(a) 110
  310.15(B)(5)(b) 113
  310.15(B)(5)(c) 114

310.15(B)(6) 118

310.15(B)(7)/Table 310.15(B)(7) [4] 118

Formulas For Adjusting Conductor Ampacity. 120
  Formula 1 120
  Formula 2 121
  Formula 3 121
    Sample Questions - Adjusting Conductor Ampacity [7] 121

Formulas For Calculating Neutral Current 126
  Examples Referencing Neutral Current [6] 127

ARTICLE 312
Cabinets, Cutout Boxes, and Meter Socket Enclosures

Wire Bending Space In Cabinets, Cutout Boxes and Meter Socket Enclosures. 129

312.6(B)(1) [3]. 131

ARTICLE 314
Outlet, Device, Pull, and Junction Boxes; Conduit Bodies; Fittings; and Handhole Enclosures

The Maximum Number Of Conductors Permitted In A Box 133

According To The NEC 133
  Boxes and Cunduit Bodies

Applying The NEC 137
  Box Fill Calculations
    When Box Contains Conductors of Same Size
      Table 314.16(A) - Metal Boxes [8] 137

    When Box Contains Conductors of Different Sizes, Devices and Other Components . . . 139
      Table 314.16(B) - Volume Allowance Required per Conductor. 143
      314.16(B) - Box Fill Calculations [12] 144

    Conduit Bodies Fill Calculations 152
      Relating To NEC 314.16(C)(1)
        314.15(C)(1) [2] 153
      Relating to NEC 314.16(C)(2). 154
        314.16(C)(2) [5] 154

The Pull Box. 156

According To The NEC 156
  Sizing Pull And Junction Boxes 156

Applying The NEC 158
  The Straight Pull
    314.28(A)(1) - Straight Pulls [7]. 161
  The Angle Pull. 165
    314.28(A)(2) - Angle Pulls or Splices [10] 166
      Considering The Diagonal Distance (DD). 167
    314.28(A)(2), Exception [2] 174
  The U-Pull 175
    314.28(A)(2) - U Pulls [3] 176
314.28(A)(2) - Angle Pull Box (when cables are installed) 177

314.28(A)(1) and (2) - Combination Pulls. 178

314.28(A)(3) - Smaller Dimensions 179

314.71(A) 179

314.71(B)(1) 179
314.71(B)(1), *Exception No. 1* 180
314.71(B)(2) 180

ARTICLE 326
Intergrated Gas Spacer Cable: Type IGS

326.116 - Conduit [2] 181

ARTICLE 330
Metal-Clad Cable: Type MC

330.24(A)(2) 182
330.24(B)
330.24(C)

ARTICLE 334
Nonmetallic-Sheathed Cable: Types NM (Romex), NMC, and NMS

334.24 - Bending Radius. 183

ARTICLE 348
Flexible Metal Conduit: Type FMC

Table 348.22 [2] 183

ARTICLE 352
Rigid Polyvinyl Chloride Conduit: Type PVC

Table 352.44 [2] 183

ARTICLE 354
Nonmetallic Underground Conduit with Conductors: Type NUCC

Table 354.24 184

ARTICLE 355
Reinforced Thermosetting Resin Conduit: Type RTRC

Table 355.44 184

ARTICLE 366 - 376 - 378
Auxiliary Gutters - Metal Wireways - Nonmetallic Wireways

According To The NEC 185
  Sizing Auxiliary Gutters and Wireways
    As applied to Auxiliary Gutters (Article 366)

    As applied to Metal Wireways (Article 376)
    and Nonmetallic Wireways (Article 378). 190

Applying The NEC 191
Gutter and Wireway Interior Cross-Sectional Area

366.22(A) and 366.22(B) [5] 194

366.23(A) [2] 198

366.23(B) 198

366.56(A) [2] 199

366.58(A) 200

366.58(B) 200

366.100(E) 201

376.22(A) and 378.22 [3] 201

376.23(A)/378.23(A) 204

376/378.23(B) [2] 204
376.56(A)/378.56 205

ARTICLE 368
Busways

368.17(B), *Exception* 206

ARTICLE 370
Cablebus

370.4(B), 370.4(C) and 370.5 [2] 207

ARTICLE 384
Strut-Type Channel Raceway

384.22 [2] 208

ARTICLE 390
Under floor Raceways

390.6 - Maximum Number of Conductors in Raceway 209

ARTICLE 392
Cable Trays

392.22(A)(1)(a) [2] 210
392.22(A)(1)(b) 212
392.22(A)(1)(c) 214
392.22(A)(3)(a) 215
392.22(A)(3)(b) 216
392.22(A)(3)(c) 216
392.22(A)(5)(b) 217

392.22(B)(1)(a) 217
392.22(B)(1)(b) 219
392.22(B)(1)(c) 219
392.22(B)(1)(d) 221

392.80(B)(2)(c) 222

Table 392.60(B) 223

ABOUT THE AUTHOR **225**

INTRODUCTION TO VOLUME 2

Volume 2 is a reflection of Chapters 1 through 3 of the National Electrical Code (NEC) that focus on a variety of selected Articles where certain portions of the featured Article require some type of an electrical calculation to be performed.

From sizing conductors and overcurrent devices supplying branch circuits, feeders and service wiring to sizing boxes or cable tray, Volume 2 is by far, an all-inclusive-one stop reference.

In considering the overall layout of Volume 2, the following color scheme was used to provide a consistent path for quick reference of selected material where:
- red identifies those articles of the NEC involving electrical calculations.
- the color green is used to identify each specific section of an article that requires some type of an electrical calculation.
- followed by the color blue which is used to identify questions that are relative to each given section.
- Answers and explanations in response to each question are identified in black.
- Where there is a cross-reference of related questions deep-purple is used.
- Finally, the color brown is used to identify NEC material, supplements, discussions and pertinent information.
- Also of important note, where sections contain multiple questions, the relative question numbers are enclosed in parenthesis ( ), followed by the total number of questions in brackets [ ]. Volume 2 also features large-print for easier visibility and reading.

Combined, Volume 2 contains over 250 challenging questions and answers along with custom illustrations to advance the understanding of the user.

ARTICLE 110 - Requirements for Electrical Installations (6)

I. General

110.14(C)(1) - Equipment Provisions (Temperature Limitations)

Before getting started, let's clear up a few things concerning equipment provisions. When the expressions "equipment provisions" or "termination provisions" of equipment are used in NEC 110.14(C)(1), the term *equipment* refers to any device such as circuit breakers, fuseholders, panelboards, load centers, disconnect switches, receptacles, light switches, etc. that has termination (connection) provisions for establishing the connection of conductors (wires) by means of mechanical pressure such as lugs, screw terminals, etc.

When the terminal (termination means) rating of such equipment is unknown, NEC 110.14(C)(1) provides conditions for associating a conductor's ampacity with the lowest terminal rating of all involved equipment.

And finally, as referenced in NEC 110.14(C) conductors with temperature ratings higher than specified for terminations shall be permitted to be used for **ampacity adjustment** (refer to NEC and Table 310.15(B)(3)(a) [Adjustment Factors for More Than Three Current-Carrying Conductors in a Raceway or Cable] *and* NEC and Table 310.15(B)(3)(c) [Ambient Temperature Adjustments for Circular Raceways Exposed to Sunlight on or Above Rooftops]), **correction** (refer to Table 310.15(B)(2)(a) [Ambient Temperature Correction Factors Based on 30°C (86°F)]), *or* **both**.

110.14(C)(1)(a) - Circuits 100 amperes or less or marked for 14 AWG through 1 AWG Conductors (1. - 3.) [3]

NEC 110.14(C)(1)(a) provides **four conditions** for determining equipment termination provisions for circuits rated 100 amperes or less *or* where equipment termination provisions are marked for the use of 1 AWG conductors or smaller (limited to 14 AWG). The conductor ampacities of Table 310.15(B)(16) is the sole reference for considering NEC 110.14(C)(1)(a) and (b).

The **first condition** per NEC 110.14(C)(1)(a)(1) reference the use of conductors rated for 60°C which limits the conductor's ampacity accordingly.

The **second condition** per NEC 110.14(C)(1)(a)(2) allows the use of a conductor with a higher temperature rating (75°C, 90°C) but limits the ampacity of such conductors to 60°C where the terminal means of a device is either rated for 60°C or unknown. For example, if the conductors of a circuit are being protected by an overcurrent device that is rated for 100A or less and the terminal rating of the overcurrent device is unknown the ampacity of the conductors must be limited to a 60°C temperature rating. Also, where a device (panelboard, disconnect switch, etc.) is marked for the use of conductors sized at 1 AWG or smaller yet the terminal rating of the device is unknown the ampacity of the conductors must also be limited to a 60°C temperature rating.

The **third condition** per NEC 110.14(C)(1)(a)(3) specifies that conductors with a higher temperature rating (75°C, 90°C) can be used if the terminal means of affiliated equipment is listed and identified for use with such conductors. When the terminal means of such equipment is rated for 75°C then the terminating conductors can bear an ampacity at 75°C. The same relationship (ampacity at 75°C) also holds true for conductors rated for 90°C. However, equipment that is listed and identified is limited for use with conductors rated up to 75°C only. Similar to the second condition, NEC 110.14(C)(1)(a)(3) also allows the use of a conductor with a higher temperature rating (90°C) yet limits the ampacity of such conductor to 75°C where the terminal means of a device is rated for 75°C. Where the terminal means of equipment reflects a dual rating (60/75°C) the equipment is listed for the use of either 60°C or 75°C conductors.

The **fourth** (and final) **condition** per NEC 110.14(C)(1)(a)(4) reference motors marked with design letters B, C, or D. Conductors (motor conductors) having an insulation rating of 75°C(167°F) or higher are allowed to be used providing the ampacity of such conductors does not exceed 75°C(167°F).

1. The terminals of a 60 amps circuit breaker are not marked to provide the terminal rating of the breaker. The breaker is used to protect a 52A non-continuous load. What size copper conductors are required to supply the load?

According to NEC 110.14(C)(1)(a)(1), when the termination provisions of equipment rated for 100 amperes or less is used, the conductors associated with the equipment must be rated for 60°C when the equipment's termination provisions are unmarked. As a minimum, 6 AWG conductors rated for 55 amps at 60°C are required to be used. Such installation (conductor's ampacity [55A] less than the rating of overcurrent device [60A]) is permitted per NEC 240.4(B). NEC 110.14(C)(1)(a)(2) allows conductors rated up to 90°C to be used however, the ampacity of such conductors must be limited to that of a conductor rated for 60°C per Table 310.15(B)(16).

2. If the circuit breaker in question No. 1. is rated for 75°C, what size conductors are required to supply the load?

According to NEC 110.14(C)(1)(a)(3), conductors with a higher temperature rating can be used if the equipment is listed and identified for use with such conductors. In this situation the 60A circuit breaker (the equipment) is considered listed and identified based on its now 75°C rating. Because the load is rated for 52 amps, 6 AWG conductors which have an ampacity of 65 amps at 75°C must be used.

3. A motor marked with a design letter B has a full-load current rating of 74A. If the motor is supplied by THW-2 copper conductors, determine the size and ampacity rating of the motor conductors.

Per NEC 110.14(C)(1)(a)(4) conductors supplying motors marked with design letters B, C, or D are allowed to have an insulation rating of 75°C(167°F) or higher providing the ampacity of such conductors does not exceed 75°C(167°F).

Referring to Table 310.15(B)(16), copper conductors bearing THW-2 insulation are rated for 90°C. Based on the 74A full-load current and the provisions of NEC 430.22, the required ampacity for the motor conductors must not be less than 92.5A (74A x 1.25). Considering the minimum ampacity at 90°C, 4 AWG THW-2 copper conductors can be used which carries a rated ampacity of 95A. However, the conditions of NEC 110.14(C)(1)(a)(4) limits the ampacity of such conductors to 75°C. At 75°C, the ampacity rating of 4 AWG THW-2 copper conductors is limited to 85A. As a result, 4 AWG THW-2 copper conductors cannot be used and must be replaced with 3 AWG THW-2 copper conductors as a minimum. A 3 AWG THW-2 copper conductor is rated for 100A at 75°C which will carry the required ampacity of 92.5A.

110.14(C)(1)(b) - Circuits rated over 100 amperes or marked for conductors larger than 1 AWG (4. - 6.) [3]

NEC 110.14(C)(1)(b) provides **two conditions** for determining equipment termination provisions for circuits rated over 100 amperes *or* where equipment termination provisions are marked for the use of conductors larger than 1 AWG (1/0 AWG up to 2000 kcmil).

The **first condition** per NEC 110.14(C)(1)(b)(1) reference the use of conductors rated for 75°C which limits the conductor's ampacity accordingly.

The **second condition** per NEC 110.14(C)(1)(b)(2) allows the use of a conductor with a higher temperature rating (90°C) but limits the ampacity of such conductor to 75°C where the terminal means of a device is either rated for 75°C or unknown. For example, if the conductors of a circuit are being protected by an overcurrent device that is greater than 100A and the terminal rating of the overcurrent device is unknown the ampacity of the conductors must be limited to a 75°C temperature rating. Also, where a device is marked for the use of conductors sized at 1/0 AWG or larger yet the terminal rating of the device is unknown the ampacity of the conductors must also be limited to a 75°C temperature rating.

4. The terminal rating of a single-phase 150A main lugs only (MLO) load center is unknown. The load center is being supplied by 2 AWG THWN-2 copper conductors. Can these conductors be used if the load center is protected by a 125A overcurrent device that's located in an upstream main load center?

The temperature (insulation) rating of a 2 AWG THWN-2 copper conductor is 90°C. Referring to Table 310.15(B)(16), the listed ampacity of the conductor is 130 amps. According to NEC 110.14(C)(1)(b)(2), conductors with a temperature rating higher (90°C) than the termination provisions of equipment (75°C) can be used, but the ampacity of the conductors must be limited to the terminal rating of the equipment. Having to operate at 75°C, limits the ampacity of the conductors to 115 amps, which is acceptable according to NEC 240.4(B).

Although, the terminal ratings of the load center and the 125A overcurrent device are unknown, the provisions of NEC 110.14(C)(1)(b)(2) are still met based on the ampere ratings of both pieces of equipment.

5. Six 4 AWG THWN-2 copper conductors are supplied from two separate 60A three-phase fusible disconnect switches which are marked for the use of conductors exceeding 1 AWG. If the conductors are ran in the same raceway which is mounted 3" above a rooftop that's exposed to a temperature of 113°F, determine the adequacy of the conductors.

NEC 110.14(C)(1)(b)(2) makes provisions for such installations. Knowing that the disconnect switches have the capacity of being fused up to 60A, the 4 AWG THWN-2 conductors (95A) can be used as is. However, because the conductors are considered current-carrying and installed in the same rooftop mounted raceway, the ampacity of the conductors must be derated based on the provisions of NEC 310.15(B)(3)(a) and (c).

Per Table 310.15(B)(2)(a) the conductors must be reduced in ampacity by 80 percent (.80) based on the number of current-carrying conductors. Because the raceway is mounted 3" above a rooftop that's exposed to a temperature of 113°F, Table 310.15(B)(3)(c) requires a temperature adder of 40°F to be applied to the given ambient temperature which totals 153°F. Referring to the correction factors of Table 310.15(B)(2)(a), at 153°F, a .58 correction factor must also be applied to the rated ampacity of the given conductors. Considering both factors, the new ampacity of the conductors is reduced to,

$$95A \times .8 \times .58 = 44.1A$$

As a result, the use of the existing 4 AWG THWN-2 copper conductors proves to be inadequate for this installation.

Based on the ratings of the 60A disconnect switches, adequate sized conductors can be determined where,

$$\frac{60A}{.8 \times .58} = 129.31A$$

The resulting calculation requires the minimum use of 2 AWG THWN-2 copper conductors (130A at 90°C) to suffice this installation. At 75°C which is the rated required ampacity per NEC 110.14(C)(1)(b)(2), the ampacity of the conductors is limited to 115A which is still adequate to meet the needs of this installation. Such installation is permitted per NEC 110.14(C).

6. If the motor in question No. 3. was protected by a 150A overcurrent device, can the same size conductors be used?

Although the terminal rating of the overcurrent device is unknown, the same 3 AWG THW-2 copper motor conductors can still be used per NEC 110.14(C)(1)(a)(4) because the conductor's ampacity is not limited to 60°C as NEC 110.14(C)(1)(a)(1) requires. Even with the overcurrent device exceeding 100A and the termination provisions unknown, the 3 AWG THW-2 conductors are not larger than 1 AWG and therefore in compliance with NEC 110.14(C)(1)(a). Remember, the overcurrent protective device of a motor is sized based on the provisions of NEC 430.52(C)(1) and Table 430.52 which allows for motor inrush current as well.

ARTICLE 210 - Branch Circuits

Article 210 covers **branch circuits**. According to Article 100, a branch circuit is defined as the circuit conductors between the final overcurrent device protecting the circuit and the outlet(s). For example, the referenced circuit conductors in simplest form could refers to an ungrounded (hot) conductor and a grounded (neutral) conductor that originates from a service panelboard or a sub-panelboard where the ungrounded conductor is connected to an overcurrent device and the grounded (neutral) conductor is connected to the grounded (neutral) bar of either panelboard where both conductors terminates in either a receptacle or lighting outlet.

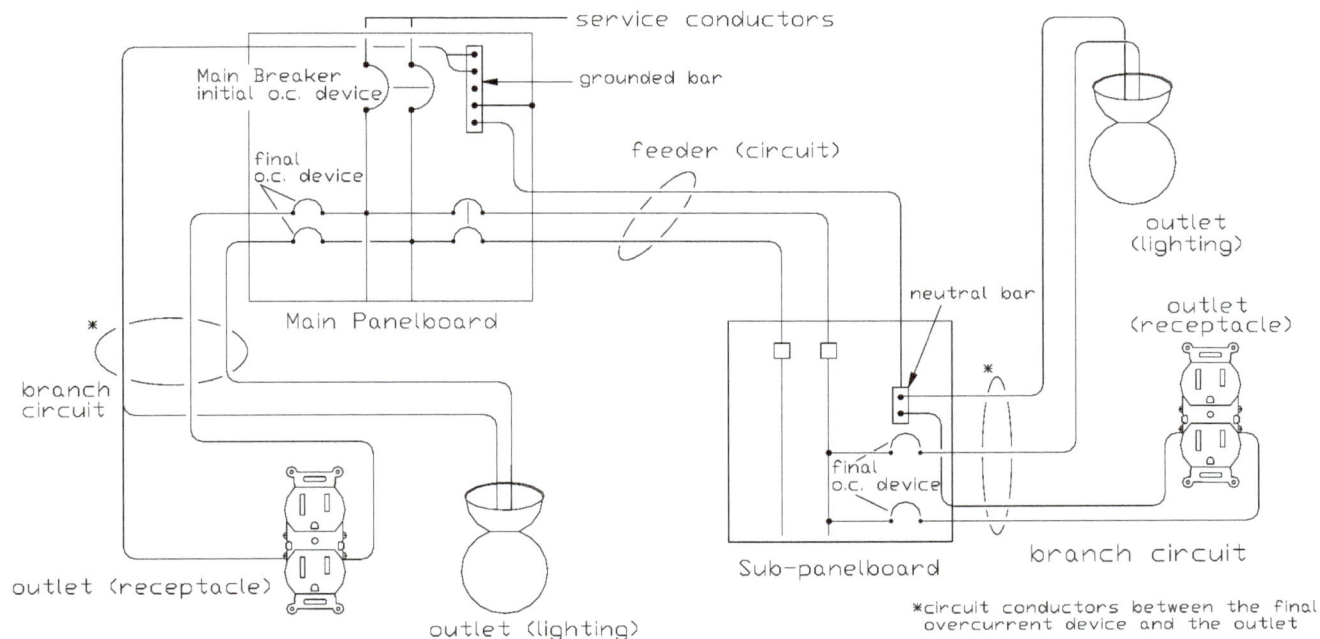

Figure 210.1 - The Branch Circuit

I. General Provisions

210.3 - Rating

1. A branch circuit consist of 6 AWG THW copper conductors that are rated for 65 amperes along with being protected by a two-pole 30 amps circuit breaker. If determining the rating of the branch circuit, which factor must be considered, the conductor's ampacity or the size of the circuit breaker?

NEC 210.3 states that the rating of a branch circuit is based on the amperes rating of the overcurrent device. Although the branch-circuit conductors are capable of supplying up to 65 amps, the ampacity of the conductors is limited to 30 amps based on the rating of the 30 amps overcurrent device (circuit breaker).

210.11(A) - Number of Branch Circuits (Branch Circuits Required)

2. Determine the minimum number of 15 or 20 amperes branch circuits required for the general lighting and receptacle loads in a 3350 square foot single-family dwelling.

The general lighting load for dwelling units is based on a unit load of 3VA per square foot according to Table 220.12. NEC 220.14(J) allows general-use receptacles to be included in the general lighting load calculations for dwelling occupancies. Therefore, the general lighting and receptacle demand load is,

$$3350 \text{ SF x } 3VA = 10{,}050 \text{ VA}$$

The demand load is now divided by 120 volts (the voltage rating of lighting and receptacle loads in a dwelling) to convert to amperes.

$$10{,}050 \text{ VA } /120V = 83.75A$$

The number of the required general lighting and receptacle branch circuits (15 or 20 amperes) can now be determined by dividing either rating into the calculated amperes value.

$$83.75A \text{ / } 15A = 5.58 \text{ (6)} \quad \text{or} \quad 83.75A \text{ / } 20A = 4.19 \text{ (5)}$$

If 15 amperes branch circuits are used 6 circuits are required or if 20 amperes branch circuits are used 5 circuits are required. Notice how each calculated number is required to be rounded up to the nearest whole number.

II. Branch-Circuit Ratings

210.19(A)(1) - General (Branch Circuits Not More Than 600 Volts) (3. - 7.) [5]

Before getting started consider the following.

NEC 210.19(A)(1) [Branch circuits], 215.2(A)(1) [Feeders] and 230.42(A)(1) [Service Conductors] requires the ampacity of a conductor to be increased by 125 percent when a conductor is being used for a continuous load. The reason for this is to keep the conductor operating as cool as possible by allowing it to function at a value that's less than its rated ampacity. For example, suppose a copper THW conductor is used to supply a 56A continuous load. By multiplying the load by 125 percent (1.25) the conductor needed for such load can be determined where,

$$56A \text{ x } 1.25 = 70A$$

Referring to Table 310.15(B)(16), a 4 AWG THW copper conductor which has an ampacity of 85 amps must be used to supply the 56 amps continuous load. Although a 6 AWG THW copper conductor has an ampacity of 65 amps which exceeds the 56 amps continuous load, using a 6

AWG conductor will only increase the ampacity of the conductor above the 56 amps continuous load by 116 percent (65A/56A = 1.16) opposed to 125 percent.

Using a 4 AWG THW copper conductor not only increases the conductor above the continuous load by 125 percent but also allows the conductor to operate at only 66 percent of its rated ampacity (56A/85A = .658 rounded up to .66). Utilizing a conductor below its rated ampacity will ensure that the conductor will operate at a cooler level thus reducing the possibilities of insulation overheating and damage.

Increasing the ampacity of a conductor by 125 percent is mostly recognized as it pertains to ungrounded (hot) conductors. However, for the grounded (neutral) conductor of a branch circuit, the same requirement holds true.

Up until the 2011 NEC, the existence of NEC 210.19(A)(1), *Exception No. 2* for branch circuits permitted grounded (neutral) conductors that were not connected to an overcurrent device to be sized at 100 percent of continuous and noncontinuous loads. As of now, the 2011 NEC only permits such *Exceptions* for feeders [NEC 215.2(A)(1), *Exception No. 2*] and service conductors [*Exception* to NEC 230.42(A)]. For more on grounded (neutral) conductors, refer to NEC 215.2(A)(2), NEC 220.61, 230.42(C) and 250.24(C)(1).

3. Branch-circuit conductors are needed to supply a 120V, 37 amps noncontinuous load. What size (THWN copper) branch-circuit conductors are required?

Because the load is noncontinuous the branch-circuit conductors can be sized based on the ampacity of the load. As a minimum, per Table 310.15(B)(16), 8 AWG THWN copper conductors that have a rated ampacity of 50 amps are required.

4. A 208V branch circuit supplies a continuous load rated for 51.7 amps. What size (RHW copper) branch-circuit conductors are required?

Branch-circuit conductors supplying continuous loads are required to be sized at 125 percent of the continuous load according to NEC 210.19(A)(1).

$$51.7 \text{ amps x } 1.25 = 64.25 \text{ amps}$$

Per Table 310.15(B)(16), as a minimum, 6 AWG RHW copper conductors which have a rated ampacity of 65 amps are required.

5. Suppose the branch-circuit conductors in question No. 4. are being protected by an overcurrent device with an unmarked terminal rating. Will the 6 AWG RHW copper conductors be sufficient to supply the load?

Since the terminal rating of the overcurrent device (equipment) is unknown and the RHW copper conductors are rated for 75°C along with the assumption that the load would be protected by an overcurrent device rated less than 100 amperes NEC 110.14(C)(1)(a)(2) states that a conductor with a higher temperature rating can be used provided the ampacity of that conductor is based on a 60°C temperature rating. At 60°C the ampacity of a 6 AWG RHW copper conductor is limited to 55 amps which is less than the required 64.25 amps calculated load. In order to suffice the calculated load a 4 AWG RHW copper conductor that's rated for 85 amps at 75°C and meets the ampacity requirement for the load at 60°C (70 amps) is minimally required.

In this case, if the terminal rating of the overcurrent device was marked, the provisions of NEC 110.14(C)(1)(a)(3) could be applied, meaning if the overcurrent device was marked for 75°C the rated ampacity (65 amps) of the 6 AWG RHW conductors could be used as is to suffice the 64.25 amps calculated load.

6. A two-wire branch circuit supplies a 5.6A continuous load along with a 5A noncontinuous load. TW copper conductors are needed to supply the loads. Determine the minimum ampacity of the conductors needed to supply the loads.

NEC 210.19(A)(1) requires the conductors to be sized at 125 percent of the continuous load and at 100 percent of the noncontinuous load. Therefore,

$$5.6A \times 1.25 + 5A = 12A$$

The conductors must have a minimum ampacity of 12 amps.

7. If the conductor's in question No. 6. were installed in a raceway with four other current-carrying conductors and the raceway was routed through an area where the ambient temperature was 37°C, determine the minimum ampacity of the conductors needed to supply the loads.

See Article 310, for related formulas 2 and 3 (located after question No. 33.)

When a branch circuit supplies a combination of continuous and noncontinuous loads, NEC 210.19(A)(1) states that the minimum branch-circuit conductor size, before the application of any adjustment or correction factors must have an allowable ampacity not less than the noncontinuous load plus 125 percent of the continuous load.

Per Table 310.15(B)(2)(a), at 37°C, type TW copper conductors have a .82 correction factor and a .80 adjustment factor per Table 310.15(B)(3)(a) based on 6 current-carrying conductors. When calculated,

$$\frac{5.6A \times 1.25 + 5A}{.82 \times .8} = 18.29A$$

As a result, the conductors must have a minimum ampacity of 18.29 amps.

For additional questions pertaining to NEC 210.19(A)(1), see Article 310, question No. 19.

210.19(A)(1) - General (Branch Circuits Not More Than 600 Volts) Informational Note No. 4

Although this informational note is not a NEC requirement the listed recommendation should be considered when sizing branch-circuit conductors.

210.19(A)(3) - Household Ranges and Cooking Appliances (Branch Circuits Not More Than 600 Volts)

Refer to Article 220 (Volume 1). See question Nos. 61. - 63. (Table 220.55, Note 4 - Branch-Circuit Load).

210.19(A)(3), *Exception No. 1* - Household Ranges and Cooking Appliances (Branch Circuits Not More Than 600 Volts)

Refer to Article 220 (Volume 1). See question No. 64. (Table 220.55, Note 4 - Branch-Circuit Load).

210.19(A)(3), *Exception No. 2* - Household Ranges and Cooking Appliances (Branch Circuits Not More Than 600 Volts)

Refer to Article 220 (Volume 1). See question No. 65 (Table 220.55, Note 4 - Branch-Circuit Load)

210.20(A) - Continuous and Noncontinuous Loads (Overcurrent Protection)

8. What size overcurrent devices are required to protect the conductors in questions Nos. 3. - 7.?

NEC 210.20(A) requires the overcurrent protection to be sized at,

100 percent for noncontinuous loads,
125 percent for continuous loads, *and* when combined
125 percent for continuous loads *plus* 100 percent for noncontinuous loads

Based on the calculations and results derived in questions 3. - 7., to be in compliance with NEC 210.20(A), for

question No. 3., a 40 amps device is required
question No. 4., a 70 amps device is required
question No. 5., a 70 amps device is required
question No. 6., a 15 amps device is required
question No. 7., a 15 amps device is required

The overcurrent protection for question No. 7. has to be determined based on the continuous and noncontinuous loads alone, that is, 5.6A x 1.25 + 5A = 12A which requires a 15 amps overcurrent device. Based on the required minimum ampacity calculated for the conductors

(18.29A), 12 AWG TW copper conductors must be used which permits a 15 amps overcurrent device per NEC 240.4(D) and NEC 240.4(D)(5) as required by NEC 210.20(B).

210.23 - Permissible Loads

9. A 120V household microwave oven appliance rated for 1600 watts will be connected to an individual branch circuit. Determine the size nonmetallic sheathed cable (Romex) and overcurrent device needed for this appliance.

DETERMINE APPLIANCE LOAD CURRENT

$$1600W / 120V = 13.33A$$

NEC 210.23 states that an individual branch circuit shall be permitted to supply any load for which it is rated meaning, the rating of the load can be up to, but not exceed, 100 percent of the ampere rating of the overcurrent protective device protecting the individual branch circuit. Based on the rated load of the microwave oven and the calculated load current per NEC 210.23, a 15 amperes individual branch circuit can be used to supply the appliance unless otherwise noted by its manufacturer, see NEC 110.3(B). In accordance with NEC 240.4(D)(3) as a minimum a 14/2 nonmetallic-sheathed cable must be used. Also based on the provisions of NEC 210.21(B)(1) a single receptacle installed on an individual branch circuit shall have an amperes rating not less than that of the branch circuit. As a result, the single receptacle that's required for this installation must be rated for 15 amperes as a minimum.

For further explanation refer to the answer given in question No. 12.

210.23(A)(1) - Cord-and-Plug Connected Equipment (Permissible Loads - 15 and 20 Ampere Branch Circuits)

10. If the microwave oven in question No. 9. was portable and connected to a multioutlet branch circuit, determine the size nonmetallic-sheathed cable (Romex) and overcurrent device needed to supply the branch circuit.

Whether realized or not NEC 210.23(A)(1) only applies to noncontinuous utilization equipment. As stated in NEC 210.23(A)(1), the rating of any one cord-and-plug-connected utilization equipment not fastened in place shall not exceed 80 percent of the branch-circuit amperes rating which is limited to either a 15 or 20 amperes branch-circuit overcurrent device per NEC 210.23(A).

The ampacity for nonmetallic-sheathed cable must be in accordance with a 60°C temperature rated conductor according to NEC 334.80. Referring to Table 310.15(B)16, 14 AWG conductors have an ampacity of 15 amps which according to NEC 240.4(D)(3) will require a 15 amps overcurrent device. Because the appliance current (load) rating cannot exceed 80 percent of the branch circuit rating per NEC 210.23(A)(1) a 15 amps overcurrent device cannot be used (13.33A /15A = .89 [89 percent]) and neither can 14 AWG conductors. At 80 percent the maximum cord-and-plug-connected load supplied by a 15A overcurrent device is limited to 12 amps (15A x .80).

A 20 amps overcurrent device can be used because the appliance current rating will not exceed the rating of the device by 80 percent (13.33A / 20A = .67). Therefore, in accordance with NEC 240.4(D)(5), 12/2 nonmetallic sheathed cable must be used. At 80 percent the maximum cord-and-plug-connected load supplied by a 20A overcurrent device is limited to 16 amps (20A x .80).

If the appliance was fastened in place it could not be connected to the multioutlet branch circuit because it would exceed the 50 percent limit imposed by NEC 210.23(A)(2).

210.23(B) - 30-Ampere Branch Circuits (11. - 12.) [2]

11. Does the 80 percent requirement apply to 25 amperes branch circuits for cord-and-plug-connected utilization equipment also?

Although 25 amperes branch circuits are not listed, the 80 percent requirement for cord-and-plug utilization equipment must also be applied in the same manner as that for a 30 amperes branch circuit. Provisions for this type branch circuit are perhaps not specified because of the infrequent use of a 25 amperes branch circuit opposed to a 30 amperes branch circuit. At 80 percent, the rating of any one cord-and-plug-connected piece of equipment would be limited to 20 amperes.

12. What size conductors and overcurrent device are required to serve a 5400W household electric dryer that's rated for 240/120V?

Often, NEC 210.23(B) is either misinterpreted or inadvertently used to size the branch-circuit conductors and overcurrent device for household electric dryers and water heaters. The concluding sentence of NEC 210.23 specifically states that the provisions of NEC 210.23(B) only applies to branch circuits supplying two or more outlets or receptacles thereby excluding individual (separate) branch circuits. Just as NEC 210.23(A)(1) reference 15 and 20 amperes branch-circuits supplying two or more outlets or receptacles where the rating of any one cord-and-plug-connected utilization equipment not fastened in place is limited to 80 percent, the same holds true for 30 amperes branch circuits according to NEC 210.23(B).

Where a 30 amperes branch-circuit supplies two or more outlets or receptacles the rating of any one cord-and-plug-connected piece of equipment is not permitted to exceed 80 percent of the branch-circuit rating or 24 amperes.

As for the household electric dryer referenced in this question, because it will require an individual branch-circuit, NEC 210.23 (second sentence) permits this type branch-circuit to supply any load for which it is rated for. This means if the circuit is rated for 30 amperes it can supply a 30 amperes load (regardless of whether hard-wired or cord-and-plug-connected), the 80 percent factor does not apply.

With a 22.5A load current (5400W/240V) a 25A overcurrent device can be used with 10 AWG branch-circuit conductors. However, if nuisance tripping occurs the overcurrent device can be increased to 30A. If the dryer was supplied by a 208V source the load current would result to 25.96A (5400W/208V) which would require a 30A overcurrent device with 10 AWG branch-circuit conductors. In either case, the provisions of NEC 110.3(B) must be adhered to along with NEC 240.4(D)(7).

Table 210.24 - Summary of Branch-Circuit Requirements

For an example and application of Table 210.24, refer to Article 220 (Volume 1). See question No. 64. (Table 220.55, Note 4 - Branch-Circuit Load)

ARTICLE 215 - Feeders

Article 215 covers **feeder circuits**. According to Article 100, a feeder is defined as all circuit conductors between the service equipment, the source of a separately derived system, or other power supply source and the final branch-circuit overcurrent device. Feeder circuit conductors originate at the service equipment where they are protected by either a circuit breaker or a set of fuses. Such circuit conductors normally terminate in sub-panelboards containing branch-circuit overcurrent devices. These branch-circuit overcurrent devices are recognized as the final overcurrent device.

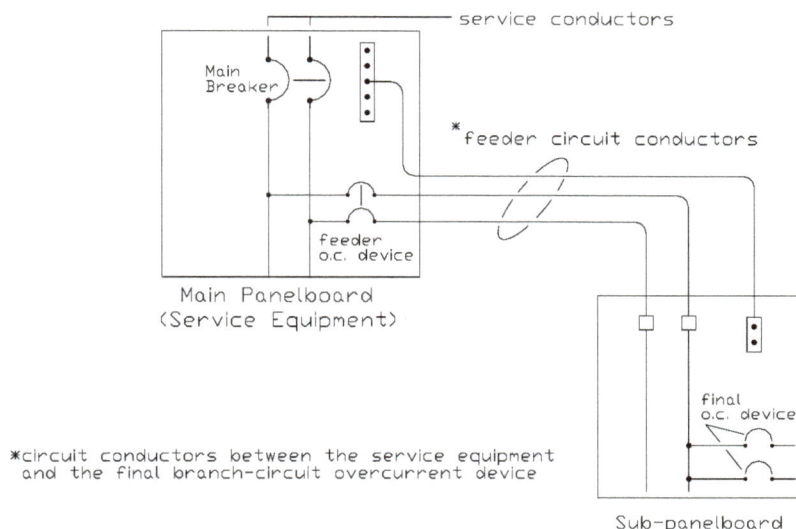

Figure 215.1 - The Feeder Circuit

215.2(A) Informational Note No. 2

Although this informational note is not a NEC requirement the listed recommendation should be considered when sizing feeder conductors.

215.2(A)(1) - General (Feeders Not More Than 600 Volts) *and* **215.3** - Overcurrent Protection
(1.- 4.) [4]

1. What size feeder conductors and overcurrent device are required to serve a 240V, 113 amps noncontinuous load, if copper conductors are required and the overcurrent device has a 75°C terminal rating?

NEC 215.2(A)(1) - noncontinuous load
The conductors must be rated for 75°C with an ampacity that is either equal to or greater than 113 amps. Per Table 310.15(B)(16), 2 AWG copper conductors are required which have a rated ampacity of 115 amps.

NEC 215.3 - noncontinuous load
A 125 amps overcurrent device can be used per NEC 240.4(B).

2. Determine the size feeder conductors, the minimum size feeder grounded conductor and the overcurrent device required to serve a three-phase, 67.6kW continuous load that's rated for 208V? Assume the use of XHHW-2 copper conductors and an unknown terminal rating of the overcurrent device.

Since the terminal rating of the overcurrent device for this installation is unknown, NEC 110.14(C)(1)(b)(1) requires the conductors be rated for 75°C with a likewise ampacity.

DETERMINE OPERATING LOAD

$$I = \frac{67,600W}{208V \times 1.732} = 187.64A$$

NEC 215.2(A)(1) and NEC 215.3 - continuous load

$$187.64A \times 1.25 = 234.55A$$

Although XHHW-2 conductors are rated for 90°C, the ampacity of the conductors must be determined at 75°C. To meet the demands of the continuous load (234.55 amps), 250 kcmil copper conductors which have a rated ampacity of 255 amps at 75°C per Table 310.15(B)(16) are required along with a 250A feeder overcurrent device per NEC 215.3. In this case, 4/0 AWG XHHW-2 copper conductors could not be used with the approval of NEC 240.4(B) because with a 230 amps rated ampacity at 75°C, the 125 percent requirement of the load would not be meet.

NEC 215.2(A)(1), *Exception No. 2* permits grounded (neutral) conductors that are not connected to an overcurrent device to be sized at 100 percent of the continuous and noncontinuous load.

NEC 215.2(A)(2) also states that the feeder circuit grounded conductor shall not be smaller than that required by NEC 250.122. Per NEC and Table 250.122 and the feeder overcurrent device being sized at 250A, as a minimum, a 4 AWG copper or a 2 AWG aluminum or copper-clad aluminum feeder grounded conductor based on the next size overcurrent device (300A) is required.

3. A 480V feeder protected by an overcurrent device with a 75°C (AL/CU) terminal rating will supply a 261 amps noncontinuous load and a 147 amps continuous load. What size overcurrent device and aluminum (75°C) conductors are required to serve the feeder loads?

According to NEC 215.2(A)(1), feeder conductors shall have an allowable ampacity not less than the noncontinuous load plus 125 percent of the continuous load. Therefore,

$$147A \times 1.25 + 261A = 444.75A$$

Based on the calculated results and Table 310.15(B)(16), as a minimum 1000 kcmil aluminum conductors rated for 445 amps at 75°C are required.

According to NEC 215.3 a feeder's overcurrent device shall not be less than 125 percent of the continuous load plus the noncontinuous load; therefore, a standard 450 amps overcurrent device is required based on the calculated results which are in compliance with NEC 240.4(B).

4. A 4W, 480/277V feeder supplying primarily nonlinear loads is installed in a metal raceway along with three other conductors that supply a three-phase motor load. The raceway will be routed and terminated in an area where the ambient temperature can reach up to 95°F. If the loads being supplied by the feeder will consist of a continuous load of 91 amps and a 58 amps noncontinuous load, what size feeder (THW copper) conductors and overcurrent device are required for this installation?

Similar to a branch circuit, when a feeder supplies a combination of continuous and noncontinuous loads that will require the ampacity of the feeder conductors to be altered, NEC 215.2(A)(1) states that the minimum feeder circuit conductor size, before the application of any adjustment or correction factors must have an allowable ampacity not less than the noncontinuous load plus 125 percent of the continuous load.

Per Table 310.15(B)(2)(a), at 95°F, type THW copper conductors have a .94 correction factor and a .70 adjustment factor per Table 310.15(B)(3)(a) based on 7 current-carrying conductors (motor and 4W feeder supplying nonlinear loads). When calculated,

$$\frac{91A \times 1.25 + 58A}{.94 \times .7} = 261.02A$$

When sized for 261 amps, 300 kcmil THW copper conductors which have a rated ampacity of 285 amps are required to serve as the feeder conductors. Considering the actual combined loads before applying the 125 percent increase (91A + 58A = 149A) the feeders conductors would only operate at approximately 52 percent (149A/285A) of its rated ampacity.

As required by NEC 215.3, the overcurrent protection must be sized no less than the noncontinuous load plus 125 percent of the continuous load.

$$91A \times 1.25 + 58A = 171.75A$$

A 175 amps overcurrent device is required based on the calculated value.

For additional questions pertaining to NEC 215.2(A)(1), see Article 310, question No. 20. For questions pertaining to motor feeder conductors and overcurrent devices refer to question Nos. 27., 28. and 46. of Article 430 (Volume 3).

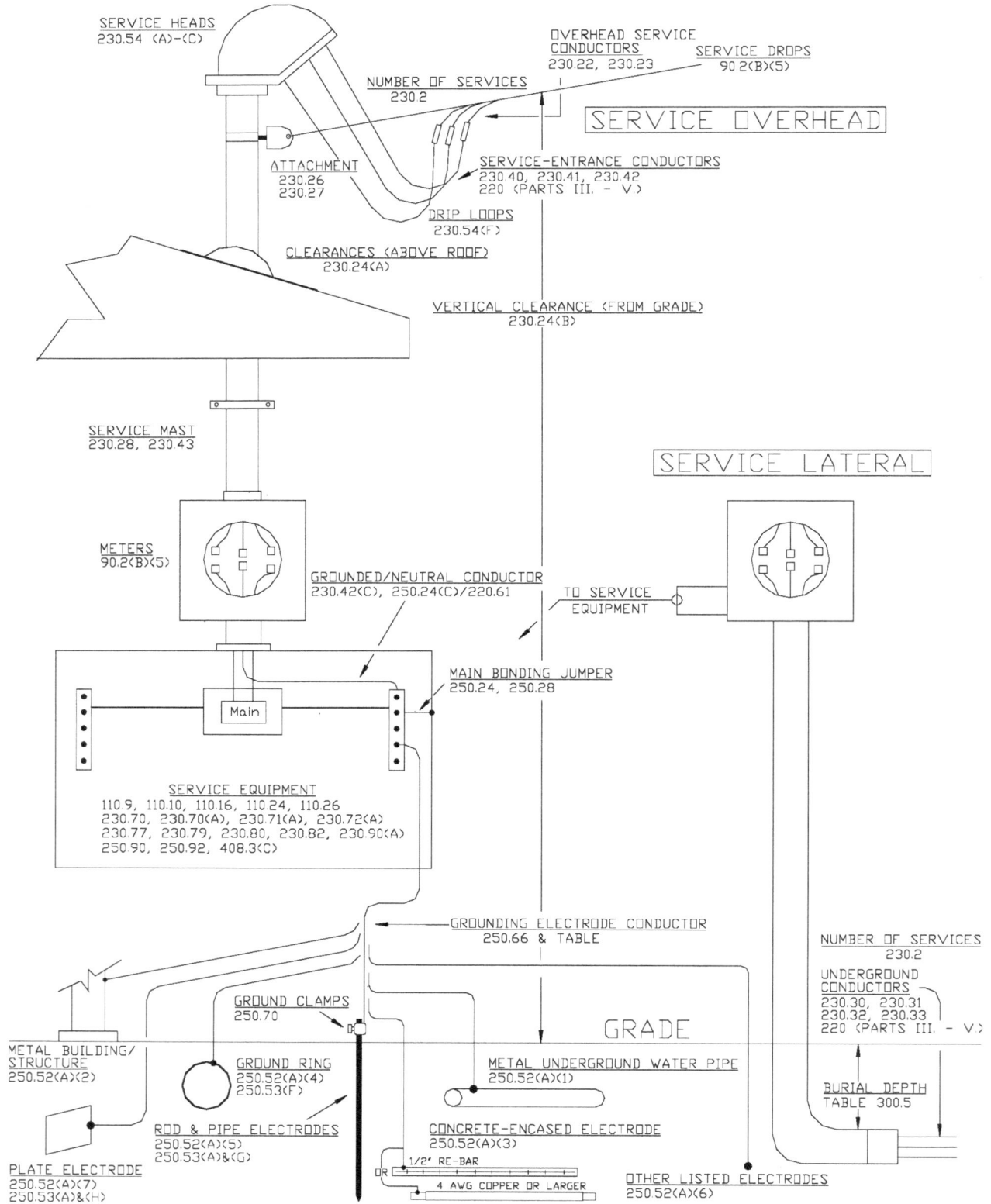

SERVICE HEADS
230.54 (A)-(C)

OVERHEAD SERVICE
CONDUCTORS
230.22, 230.23

SERVICE DROPS
90.2(B)(5)

NUMBER OF SERVICES
230.2

SERVICE OVERHEAD

ATTACHMENT
230.26
230.27

SERVICE-ENTRANCE CONDUCTORS
230.40, 230.41, 230.42
220 (PARTS III. - V.)

DRIP LOOPS
230.54(F)

CLEARANCES (ABOVE ROOF)
230.24(A)

VERTICAL CLEARANCE (FROM GRADE)
230.24(B)

SERVICE MAST
230.28, 230.43

SERVICE LATERAL

METERS
90.2(B)(5)

GROUNDED/NEUTRAL CONDUCTOR
230.42(C), 250.24(C)/220.61

TO SERVICE
EQUIPMENT

MAIN BONDING JUMPER
250.24, 250.28

Main

SERVICE EQUIPMENT
110.9, 110.10, 110.16, 110.24, 110.26
230.70, 230.70(A), 230.71(A), 230.72(A)
230.77, 230.79, 230.80, 230.82, 230.90(A)
250.90, 250.92, 408.3(C)

GROUNDING ELECTRODE CONDUCTOR
250.66 & TABLE

NUMBER OF SERVICES
230.2

UNDERGROUND
CONDUCTORS
230.30, 230.31
230.32, 230.33
220 (PARTS III. - V.)

GROUND CLAMPS
250.70

GRADE

METAL BUILDING/
STRUCTURE
250.52(A)(2)

GROUND RING
250.52(A)(4)
250.53(F)

METAL UNDERGROUND WATER PIPE
250.52(A)(1)

BURIAL DEPTH
TABLE 300.5

ROD & PIPE ELECTRODES
250.52(A)(5)
250.53(A)&(G)

CONCRETE-ENCASED ELECTRODE
250.52(A)(3)

PLATE ELECTRODE
250.52(A)(7)
250.53(A)&(H)

OR 1/2" RE-BAR

4 AWG COPPER OR LARGER

OTHER LISTED ELECTRODES
250.52(A)(6)

ARTICLE 230 - Services

230.42(A) - General Minimum Size and Rating (1. - 2.) [2]

1. Type USE aluminum conductors are being used to supply a 3845 sq. ft. house. The calculated load (215 amps) for this house was based on the use of a 240/120V single-phase service. What size service and ungrounded service conductors are needed for this dwelling unit?

As referenced in NEC 230.42(A) and permitted by Table 310.15(B)(7), a 225 amps service is needed which requires the use of 250 kcmil aluminum service conductors.

2. A three-phase, 4W-208/120V service supplies an 8875 square feet convenient store. The store's lighting load consist of 7 - 208V single-phase parking light fixtures totaling 23 amps and 2 - 208V single-phase 1800W commercial signs in addition to the interior lighting. The store is equipped with 10 - 120V, 2.5A electronic cash registers, 77 - 120V receptacles, 3 - 208V three-phase 15.5kVA air conditioning loads and a variety of other noncontinuous loads totaling 45.2kVA where 28 percent of these loads are neutral loads. What size THWN ungrounded copper service conductors are required to supply the store?

The store's demand load must first be determined based on the requirements of NEC 230.42(A).

CONTINUOUS LOADS
General Lighting - 8875 SF x 3VA*   = 26,625VA
Outdoor Lights - 208V x 23A   = 4,784VA
Signs - 2 x 1800VA   = 3,600VA
Cash Registers - 120V x 2.5A x 10   = 3,000VA
* Table 220.12   38,009VA

$$38,009VA \times 1.25 = 47,511.25VA$$

NONCONTINUOUS LOADS - Receptacles (Table 220.44) - 180VA x 77 = 13,860VA
10,000VA (@100%) = 10,000VA
3,860VA (@50%)   = 1,930VA
                    11,930VA

AC - 3 x 15,500VA = 46,500VA
Others - 45,200VA x .72 = 32,544VA

Total - 11,930VA + 46,500VA + 32,544VA = 90,974VA

DEMAND LOAD - 47,511.25VA + 90,974VA = 138,485.25VA

THREE-PHASE DEMAND LOAD

$$\frac{138,485.25VA}{208V \times 1.732} = 384.41A$$

Based on the calculation per Table 310.15(B)(16), 3 - 600 kcmil THWN copper conductors are required to serve as the ungrounded conductors to supply the store. The conductors are rated for 420A.

230.42(C) - Grounded Conductors (Minimum Size)

3. What size neutral (grounded) THWN copper conductors are needed for the service in question No. 2?

The service neutral (grounded) conductors must be sized based on applicable loads per Article 220 and the demand factors of NEC 220.61, where applicable. The *Exception* to NEC 230.42(A)(1) permits grounded (neutral) conductors that are not connected to an overcurrent device to be sized at 100 percent of the continuous and noncontinuous loads.

LOADS

General Lighting	-	26,625VA
Cash Registers	-	3,000VA
Receptacles	-	11,930VA
Others	-	12,656VA (28% of 45,200VA)
		54,211VA

NEUTRAL DEMAND LOAD = 54,211VA

$$\frac{54,211VA}{208V \times 1.732} = 150.48A$$

Based on the calculation, a 2/0 AWG copper conductor per Table 310.15(B)16 is required to serve as the service neutral conductor which is in compliance with NEC 230.42(C) and 250.24(C)(1). The conductors are rated for 175A.

230.79(C) - One Family Dwelling (Service Equipment)

4. The demand load for a small 525 SF single-family residence is calculated at 57 amps. Can a 60 amps service disconnecting means be used for this residence?

According to NEC 230.79(C), a 3-wire, 100 amps service disconnecting means is the smallest rating allowed for a one-family dwelling.

230.90(A) - Overcurrent Protection (Service Equipment) (5. - 6.) [2]

5. The incoming service for a saw mill has a 3φ-240/120V, 4W service. Although the calculated 287 amps demand load for the mill is far less than the 400A three-phase panelboard (with main breaker) that's being installed as the main service, determine the size line conductors required for this installation if THWN copper conductors are used.

NEC 230.90(A), *Exception 2* permits the use of 500 kcmil THWN copper conductors which have a rated ampacity of 380 amps.

6. A 200A panelboard with a main circuit breaker is being installed along with 3 - 150 amps fusible disconnect switches as the service equipment for a new restaurant. The calculated service demand load for this installation is 432 amps. Determine the minimum size 75°C copper service conductors needed.

NEC 230.90(A), *Exception 3* permits the sum of the ratings of the circuit breaker and fuses (650A) to exceed the ampacity of the service conductors, providing the calculated load (432A) does not exceed the ampacity of the service conductors.

Being the case, 700 kcmil copper conductors with a rated ampacity of 460 amps at 75°C can be used.

```
OVERCURRENT PROTECTION REFERENCES
            per ARTICLES


ARTICLE  TITLE                              REFERENCE
210      Branch Circuits                    210.20
215      Feeders                            215.3
230      Services                           230.79
240      Overcurrent Protection             240.4(B)-(D), 240.6, 240.21
250      Grounding & Bonding                Table 250.122
310      Conductors for Gen. Wiring         Table 310.15(B)(7)
366      Auxiliary Gutters                  366.56(D)
368      Busways                            368.17
370      Cablebus                           370.5
392      Cable Trays                        Table 392.60(A)
400      Flexible Cords and Cables          400.13
402      Flexible Wires                     402.12
408      Switchboards and Panelboards       408.36
409      Industrial Control Panels          409.21(C)
422      Appliances                         422.11, 422.13
424      Fixed Electric Space-Htg Equip.    424.3(B), 424.22
                                            424.72, 424.82
426      Fixed Outdoor Elect. Deicing ....  426.4
427      Fixed Elect. Heating Equip. ....... 427.4
430      Motors, Motor Circuits, and .....  430.52(C)&(D), 430.62(A)&(B)
                                            430.63, 430.72
440      Air-Cond. and Refrig. Equip.       440.22(A)&(B)
445      Generators                         445.12
450      Transformers and Tran. Vaults      Table 450.3(A)&(B), 450.4(A)
455      Phase Converters                   455.7(A)&(B)
460      Capacitors                         460.8(B)
517      Health Care Facilities             517.73(A)&(B)
522      Control Systems for Perm. .....    522.23
530      Motion Picture and Tele. .......... 530.18
550      Mobile Homes, Manufactured ....    550.11(A)-(C), 550.32(C)
                                            550.33(B)
551      Recreational Vehicles and .......  551.43(A), 551.74
552      Park Trailers                      552.10(E)
590      Temporary Installations            590.4(A)-(C)
600      Electric Signs and Outline .....   600.5(A)&(B)
610      Cranes and Hoists                  610.41(A)&(B), 610.42(A)&(B)
620      Elevators, Dumbwaiters, .........  620.61(A)-(C)
625      Electric Vehicle Charging .......  625.21
630      Electric Welders                   630.12, 630.32
660      X-Ray Equipment                    660.6(A)&(B)
665      Induction and Dielect. Htg. ....   665.11
668      Electrolytic Cells                 668.3(C)(2)
669      Electroplating                     669.9
670      Industrial Machinery               670.4(C)
675      Electrically Driven or Cont. ..    675.7(A)&(B), 675.22(A)&(B)
690      Solar Photovoltaic (PVC) .......   690.6(E), 690.9(A)-(E)
                                            690.10(B)
692      Fuel Cell Systems                  692.9
694      Small Wind Electric Syst. .......  694.15(A)-(C)
695      Fire Pumps                         695.4(B)(2), 695.5(B)&(C)(2)
700      Emergency Systems                  700.25 - 700.27
701      Legally Required Standby ......    701.25
705      Interconnected Electric .........  705.10(D)(2), 705.30
                                            705.65(A)&(B), 705.130
708      Critical Operations Power ......   708.50 - 708.54
725      Class 1, Class 2, and Class ..    725.45(A)(E)
727      Instrumentation Tray Cable        727.9
760      Fire Alarm Systems                 760.43
```

ARTICLE 240 - Overcurrent Protection

In NEC 110.14(C) and (C)(1) of Article 110 the size and altered size conductors needed for making electrical terminations to equipment per temperature ratings was the primary focus. In Articles 210, 215 and 230 requirements are given for adjusting the size of conductors when supplying continuous and noncontinuous loads. Along with the mentioned articles, Article 310 contains a variation of tables reflecting ampacities per conductor sizes along with supplemental provisions for making ampacity corrections and adjustments where needed for conductor sizing.

Although the scope of Article 240 covers the general requirements for overcurrent protection and overcurrent protective devices, when it comes to sizing conductors per use of overcurrent protective devices, Article 240 is often perceived as the most challenging of all mentioned articles. For this reason much effort was exerted in the preparation of this topic to ensure the reader of a clear understanding of all inclusive material pertaining to Article 240.

Part I (General)

240.3 - Other Articles

1. Explain the conditions of NEC 240.3 and its related Table.

Because Article 240 only covers the general provisions for protecting conductors and equipment, NEC 240.3 clearly makes known the use of other applicable articles that provides specific guidelines for protecting selected types of equipment. Table 240.3 reference the use of 35 articles that pertains to the use and application of overcurrent protection for various types of equipment.

Explain the conditions of NEC 240.4.

According to NEC 240.4, conductors other than flexible cords, flexible cables, and fixture wires shall be protected against overcurrent in accordance with their ampacities specified in NEC 310.15 unless otherwise permitted or required in 240.4(A) through (G). As a result, NEC 240.4 provides further allowance or alternative means for the protection against conductor overcurrent per NEC 240.4(A) through (G). Needless to say, without these provisions, the overcurrent device protecting a conductor would be required to be rated the same or less than the ampacity of a conductor.

240.4(B) - Overcurrent Devices Rated 800 Amperes or Less (Protection of Conductors)
(2. - 7.) [6]

Of the three conditions pertaining to NEC 240.4(B), condition **(1)** could appear or come off as being out of place compared to conditions **(2)** and **(3)**. With conditions **(2)** and **(3)** the instructions are more obvious and better understood. For this reason, clarifying the provisions of condition **(1)** will be made more apparent through the following example.

> A 20 amperes (A) branch circuit supplying 12 receptacles will be added to the exterior walls of a building to provide a 120V source for various non-continuous portable loads. What size THW copper conductors are required to supply the receptacles?

According to NEC 220.14(I), receptacle outlets shall be calculated at not less than 180 volt-amperes (VA) for each single or multiple receptacle on one yoke. As a result, the branch-circuit load would amount to, 180VA x 12 (receptacles) = 2160VA *where* 2160VA/120V = 18A. Therefore requiring (based on the rating of the 20A branch circuit) per NEC 240.4(D)(5) and Table 310.15(B)(16), the use of 12 AWG THW conductors which have a rated ampacity of 25A, which exceeds the calculated branch-circuit load and the rating of the branch circuit. Although the conductors will be a part of the installation described in NEC 240.4(B)(1), it does not violate this condition.

If the conductors (2-wire) supplying the receptacles were enclosed in a raceway with four other current-carrying conductors along with an equipment grounding conductor and exposed to an ambient temperature of 93°F,what size conductors would now be required?

Since the ambient temperature will be higher than 86°F (30°C) and more than three current-carrying conductors will be enclosed in the same raceway, the ambient temperature correction factor [Table 310.15(B)(2)(a)] and the adjustment factor [Table 310.15(B)(3)(a)] must be applied to alter the ampacity of the required conductors.

First considering the use of the 12 AWG THW copper conductors based on their related 75°C temperature rating, a correction factor of .94 and a .80 adjustment factor must be applied towards the ampacity of the conductors. Therefore, 25A x .94 x .80 = 18.8A.

At 18.8A, the 12 AWG conductors' new ampacity would now fall below the rating of the 20A protective device which is used to protect these conductors that supply the receptacles. By inversing the application of the previous calculation, using the ampacity of the 12 AWG conductors the results are as so, 25A / (.94 x .80) = 33.24A, which implies the use of conductors that have an ampacity that is either equal to or exceed 33.24A. Based on this ampacity, as a minimum 10 AWG THW copper conductors are required which has a rated ampacity of 35 amperes at 75°C. If the rated ampacity of the 10 AWG conductors was reduced based on the correction and adjustment factors the ampacity would still exceed the rating of the 20A overcurrent device thus remaining in compliance with NEC 240.4(B)(1). Observe, 35A x .94 x .80 = 26.32A. With this example, condition **(1)** now becomes more apparent.

2. A 265A noncontinuous load is protected by a 300A circuit breaker. Determine the minimum size THWN copper and aluminum conductors required to serve this load. What size overcurrent device and conductors would be required *if*, **(a)** the load was continuous, **(b)** exposed to an ambient temperature of 95°F, **(c)** consist of seven current-carrying conductors, **(d)** all factors were considered? Assume this load will not supply any type receptacle branch circuits.

Per Table 310.15(B)(16) - as a minimum, based on the use of copper, 300 kcmil THWN conductors rated for 285 amps are needed whereas, based on the use of aluminum, 400 kcmil THWN conductors rated for 270 amps are required. The use of either conductor exceeds the rating of the load.

(a) *Continuous Load* - Load increased by 125 percent - 265A x 1.25 = 331.25A

Considering the calculated load, per Table 310.15(B)(16) - as a minimum, based on the use of copper, 400 kcmil THWN conductors rated for 335 amps are needed whereas, based on the use of aluminum, 600 kcmil THWN conductors rated for 340 amps are needed. Based upon the use of either conductor, per NEC 240.4(B) a 350A overcurrent device is permitted.

(b) *Exposed to an ambient temperature of 95°F* - Per Table 310.15(B)(2)(a) - based on the 95°F ambient temperature, the correction factor at 75°F is .94. The ampacity to determine the size of the needed conductors requires adjusting to compensate for the ambient temperature. This results to the load being divided by the correction factor. Therefore, 265A/.94 = 281.91A. At 281.91A, per Table 310.15(B)(16), as a minimum, 300 kcmil THWN copper conductors can also be used based on the required ampacity to compensate for the 95°F ambient temperature. The same holds true for the use of the 300A overcurrent device to protect the conductors. As a minimum for aluminum conductors, 500 kcmil THWN conductors rated for 310A are now needed. Per NEC 240.4(B) a 350A overcurrent device is permitted.

(c) *Seven current-carrying conductors* - Per Table 310.15(B)(3)(a) - based on seven current-carrying conductors the adjustment factor is .70. The ampacity to determine the size of the needed conductors now requires adjusting to compensate for the current-carrying conductors. This results to the load being divided by the adjustment factor. Therefore, 265A/.70 = 378.57A. At 378.57A, per Table 310.15(B)(16), as a minimum, 500 kcmil THWN copper conductors rated for 380A are required based on the calculated ampacity to compensate for the current-carrying conductors. Per NEC 240.4(B) a 400A overcurrent device is permitted. As a minimum for aluminum conductors, 750 kcmil THWN conductors rated for 385A are now required. Per NEC 240.4(B) a 400A overcurrent device is permitted.

(d) *All factors considered* (continuous load, ambient temperature and current-carrying conductors) results to, 265A x 1.25/.94 x .70 = 503.42A. At 503.42A, per Table 310.15(B)(16), as a minimum, 900 kcmil THWN copper conductors rated for 520A are required based on the ampacity needed to compensate for all factors. Per NEC 240.4(B), a 600A overcurrent device is permitted. As for aluminum conductors, 1500 kcmil THWN conductors rated for 520A are now required. Per NEC 240.4(B), a 600A overcurrent device is also permitted.

3. Size 2 AWG THWN copper conductors are used to supply a load rated less than the ampacity of the conductors. What size overcurrent protection is required to protect the 2 AWG conductors?

Per Table 310.15(B)(16), 2 AWG THWN copper conductors are rated for 115A, therefore the use of a 125A overcurrent protection is permitted.

4. It is determined that the electrical service for a new building will require 182 amps. What size 75°C copper conductors and overcurrent device are required, if the load is continuous?

Reference NEC 230.42(A)(1), 230.90(A), *Exception 2* and Table 310.15(B)(16).

$$182A \times 1.25 = 227.5A$$

Use 4/0 copper conductors (rated for 230 amps) and a 250A overcurrent device.

5. Two sets of 4/0 AWG parallel copper conductors will terminate on the lugs of a three-phase 600A fusible disconnect switch rated for 75°C. What size fuses are required to protect the conductors?

A 4/0 AWG conductor rated for 75°C [Table 310.15(B)(16)] has an ampacity of 230 amps. Therefore,

$$230 \text{ amps x } 2 = 460 \text{ amps}$$

Based on the total ampacity of the parallel conductors, the use of three 500A fuses are permitted.

6. A feeder protected by an 800A overcurrent device will supply a continuous 615 amps load. What size 75°C aluminum parallel conductors are required to feed the load?

Reference NEC 215.2(A)(1), 215.3 and Table 310.15(B)(16). According to NEC 215.2(A)(1), feeder conductors shall have an allowable ampacity not less than 125 percent of a continuous load therefore,

$$615A \text{ x } 1.25 = 768.75A$$

Based on the results of the calculation, the 800A overcurrent device exceeds the continuous load by 125 percent as required by NEC 215.3.

To determine the size parallel conductors where 2 conductors are used, the size of the given overcurrent protection is considered based on the provisions of NEC 240.4(B) where,

$$800A / 2 = 400A$$

Because NEC 240.4(B) permits the next higher standard overcurrent device rating above the ampacity of the conductors being protected, conductors rated less than 400A can be used.

Therefore, two 750 kcmil aluminum conductors rated for 385 amps each at 75°C can be used because they both will serve to satisfy the calculated continuous load (768.75A / 2 = 384.38A). Now if the use of three parallel conductors is desired where 800A / 3 = 266.67A, then three 400 kcmil aluminum conductors rated for 270 amps each at 75°C could be used which would also satisfy the calculated continuous load (768.75 / 3 = 256.25A). If three 350 kcmil aluminum conductors at 250A each were used, the total ampacity of the conductors (250A x 3 = 750A) would be less than the 768.75A continuous load. (400 kcmil conductors are required).

7. If used in a dwelling where the incoming service is rated for 240/120V-1ϕ. What size THWN copper conductors are required to supply a 150 amps load center with a main breaker?

Referring to Table 310.15(B)(7), a 1 AWG THWN copper conductor is permitted to supply the load center. Although contrary to the condition of NEC 240.4(B)(2) which states, "the ampacity of the conductor(s) does not correspond with the standard amperes rating of a fuse or circuit breaker" where in this application a 1/0 AWG THWN conductor rated for 150 amps does, the provisions of NEC and Table 310.15(B)(7) supersedes NEC 240.4(B). Bear in mind that if this application was a non-dwelling or a dwelling where the incoming service was rated for 208/120V single-phase the provisions of Table 310.15(B)(7) would not apply and a 1/0 AWG THWN copper conductor would be required to supply the 150A load center.

240.4(C) - Overcurrent Devices Rated over 800 Amperes (Protection of Conductors) (8. - 10.) [3]

8. What size overcurrent protection is required to protect three sets of 250 kcmil THW parallel copper conductors which originate in a 3-phase service panelboard that's rated for 1200A?

According to Table 310.15(B)(16), a 250 kcmil THW copper conductor has an ampacity of 255 amps where, 255 amps x 3 = 765 amps.

An 800A overcurrent device is required. An overcurrent device larger than 800 amps would require the ampacity of the parallel conductors to be either equivalent or greater than the rating of the overcurrent device. Because the service panelboard is rated for 1200A, an overcurrent protective device up to 1200A can be used.

If a 1000A overcurrent device was used, what size conductors would be required?

Now, according to NEC 240.4(C), where the overcurrent device is rated over 800 amperes, the ampacity of the conductors it protects shall be equal to or greater than the rating of the overcurrent device defined in NEC 240.6(A). Now considering the use of three parallel conductors where 1000A / 3 = 333.33A.

Three 400 kcmil THW copper conductors, rated for 335 amps each at 75°C as a minimum are required to be used to satisfy the provisions of NEC 240.4(C) where the total ampacity of the conductors at 1005A (335A x 3) exceeds the rating of the 1000A overcurrent device.

9. A three-phase service rated for 480/277V supplies a local restaurant. The service consist of two three-phase 800A fusible disconnect switches fused at 600A (601A) where the terminals of the disconnects are rated for 75°C. What size aluminum service conductors are required if four parallel conductors are installed per phase?

Although, the total service rating has the potential of reaching up to 1600A based on the two 800A fusible disconnect switches, the service is limited to 1200A because both switches are fused at 600A. Because each disconnect switch has the potential of being fused up to and not exceeding 800A, each separate application still falls under the provision of NEC 240.4(B) (Overcurrent Devices Rated 800 Amperes or Less). Therefore, at 800A and the use of 4 parallel conductors per phase - per disconnect switch, the parallel conductors can have a combined ampacity that is less than 800A. As a minimum, four 4/0 AWG aluminum conductors rated for 75°C at 180A each can be used where, 180 amps x 4 = 720 amps. At maximum, the applied load cannot exceed the combined ampacity of the aluminum conductors.

10. Reconsider question No. 9. If the service is now a 1600A fusible disconnect switch fused at either 1200A or 1600A where 4 sets of parallel conductors are used.

Considering the use of 75°C rated aluminum conductors where 1200 amps / 4 = 300 amps if 1200A fuses are used.

For this application, as a minimum, 500 kcmil aluminum conductors are required where the total ampacity of four such conductors is 1240 amps (310A x 4) which exceed the rating of the 1200A fuses per NEC 240.4(C).

Now, considering the use of 75°C rated aluminum conductors where 1600 amps / 4 = 400 amps if 1600A fuses are used.

For this application, as a minimum, 900 kcmil aluminum conductors are required where the total ampacity of four such conductors is 1700 amps (425A x 4) which exceed the rating of the 1600A fuses per NEC 240.4(C).

240.4(D) - Small Conductors (Protection of Conductors)

11. What size overcurrent protection is required to protect 14 AWG, 12 AWG and 10 AWG copper conductors? What size overcurrent protection is required to protect 12 AWG and 10 AWG aluminum and copper-clad aluminum conductors?

The 2008 edition of the National Electrical Code has taken additional measures to individualized NEC 240.4(D) based on conductor size and material type. Although NEC 240.4(D) has been made more distinct the provisions does not apply to all applications where 14 AWG, 12 AWG and 10 AWG conductors requires overcurrent protection.

For applications pertaining to protecting copper conductors

According to NEC 240.4(D)(3) the only size overcurrent device that is permitted to protect a 14 AWG copper conductor must be rated no more than 15A.

According to NEC 240.4(D)(5) the maximum size overcurrent device that is permitted to protect a 12 AWG copper conductor must be rated no more than 20A. As a minimum, a 15A overcurrent device can also be used to protect a 12 AWG copper conductor. The rule to remember is, the overcurrent device cannot exceed 20A.

According to NEC 240.4(D)(7) the maximum size overcurrent device that is permitted to protect a 10 AWG copper conductor must be rated no more than 30A. As a minimum, a 15A, 20A or 25A overcurrent device can also be used to protect a 10 AWG copper conductor. The rule to remember is, the overcurrent device cannot exceed 30A.

For applications pertaining to protecting aluminum and copper-clad aluminum conductors

According to NEC 240.4(D)(4) the only size overcurrent device that is permitted to protect a 12 AWG aluminum and copper-clad aluminum conductor must be rated no more than 15A.

According to NEC 240.4(D)(6) the maximum size overcurrent device that is permitted to protect a 10 AWG aluminum and copper-clad aluminum conductor must be rated no more than 25A. As a minimum, a 15A or 20A overcurrent device can also be used to protect a 10 AWG aluminum and copper-clad aluminum conductor. The rule to remember is, the overcurrent device cannot exceed 25A.

NEC 240.4(D) specifically states that the required conductor/overcurrent device relation must still be maintained regardless of correction (ambient temperature) or adjustment (number of current-carrying conductors) factors.

Also, great emphasis must be placed on the fact that the conductor/overcurrent device relation per NEC 240.4(D) is not applicable for every application in the NEC. Where specific conductor/overcurrent device applications are required such as for tap conductors, motor conductors, etc., NEC 240.4(E), 240.4(G) and Table 240.4(G) must be referenced and adhered to accordingly.

240.4(E) - Tap Conductors

NEC 240.4(E) is another section of Article 240 which allows the ampacity of a conductor to be rated less than the overcurrent protective device. Listed below are sub-sections that are applicable to NEC 240.4(E).

240.4(E)(1) - (Tap Conductors) 210.19(A)(3), Household Ranges and Cooking Appliances

Refer to Article 220 of Volume 1. See question Nos. 61. - 65. (Table 220.55, Note 4 - Branch-Circuit Load).

240.4(E)(4) - (Tap Conductors) 368.17(B), Reduction in Ampacity Size of Busway

See question No. 1. of Article 368.

240.4(E)(6) - (Tap Conductors) 430.53(D), Single Motor Taps

Although there are no questions pertaining to this sub-section as it relates to branch-circuit single motors taps, see question No. 43. of Article 430 for feeder taps.

240.4(G) - Overcurrent Protection for Specific Conductor Applications

12. If 12 AWG conductors were used to supply a single-phase motor would a 20A overcurrent device be required to protect the conductors?

According to Table 240.4(G), when overcurrent protection is required to protect conductors supplying motor applications, Part III, IV, V, VI or VII of Article 430 must be referenced to determine the proper size overcurrent device needed. In this case NEC 430.52(C)(1) of Part IV of Article 430 is applied along with the provisions of Table 430.52 for determining the overcurrent protection for motor conductors. The use of Part IV of Article 430 for determining such requirements and others is referenced in NEC 430.51. As for the use of a 20A overcurrent device several factors must first be considered, primarily the type motor and the desired overcurrent device per Table 430.52, to name a few. As you can see, the procedure for sizing the overcurrent protection for the 12 AWG motor conductors requires more than just simply referring to NEC 240.4(D). When sizing overcurrent protection for motors, the possibilities of using an overcurrent device that greatly exceeds the ampacity of a motor's conductors will always exist. This is only one reason why the use of NEC 240.4(D) cannot be applied for all applications involving 14 AWG, 12 AWG and 10 AWG conductors.

Part II (Location)

Tap Conductors

Before taking on questions pertaining to tap conductors, let's take a look at the definition provided in NEC 240.2 for that of *tap conductors* by first defining and illustrating the term *tap*. Unlike a splice where two or more conductors of the same ampacity are connected together and are both protected by the same rated overcurrent device, a *tap* is where a conductor having a lesser ampacity is connected to a conductor having a larger ampacity and where proper overcurrent protection is provided for only the conductor of the larger ampacity.

According to the National Electrical Code, a *tap conductor* is defined per NEC 240.2 as a conductor, *other than a service conductor* (see Figure 240.1), that has overcurrent protection ahead of its point of supply that exceeds the value permitted for similar conductors that are protected as described elsewhere in NEC 240.4.

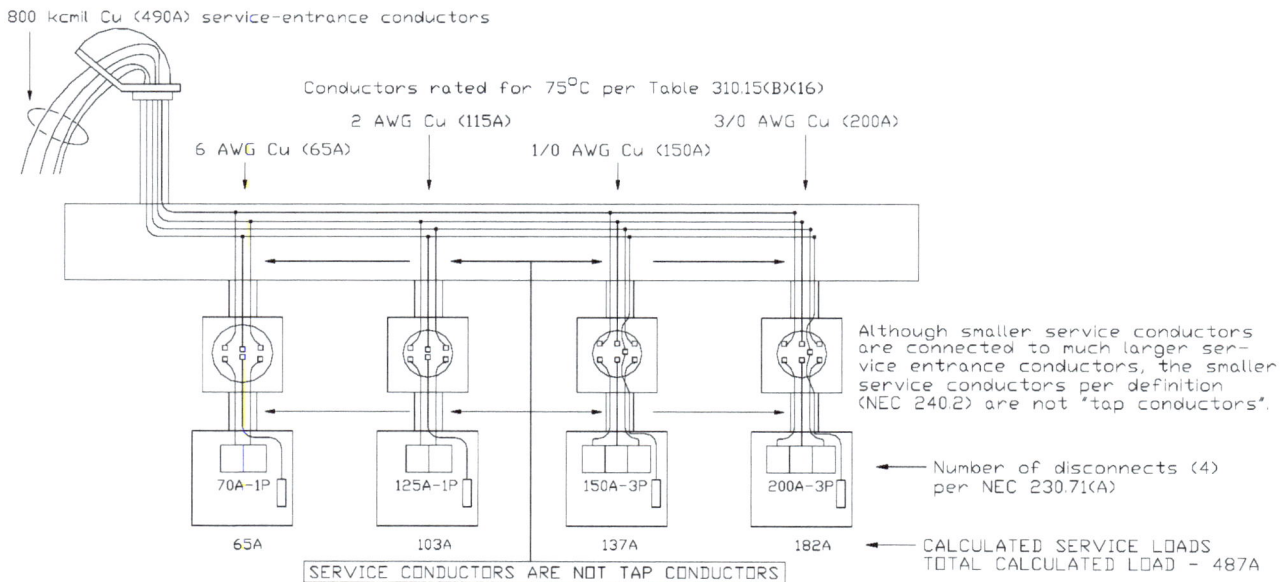

Figure 240.1 - Other than a service conductor

Having an understanding of *"what a tap conductor is"*, is very important because of the difficulties most users encounter when attempting to interpret and apply sub-sections (B) [Feeder Taps] and (C) [Transformer Secondary Conductors] of NEC 240.21. Both sub-sections are an *Exception* to NEC 240.21. For simplicity, just remember that the point of supply is where the actual connection (tap) is made between a supply (feeder) conductor and a conductor that has an ampacity less than the supply conductor. If a conductor having a specific ampacity was connected to another conductor of the same ampacity, the connection then at the point of supply is called a *splice*, you see the difference. Also, simply because a smaller conductor is connected to a larger conductor, the smaller conductor in some applications *is not always* considered a tap conductor. Neither can it be said in such applications that the smaller conductor is tapped to the larger conductor. In summary, a tap conductor is either inadequately protected based upon its rated ampacity or insufficiently protected at its point of supply. For practical examples, see Figures 240.2 and 3.

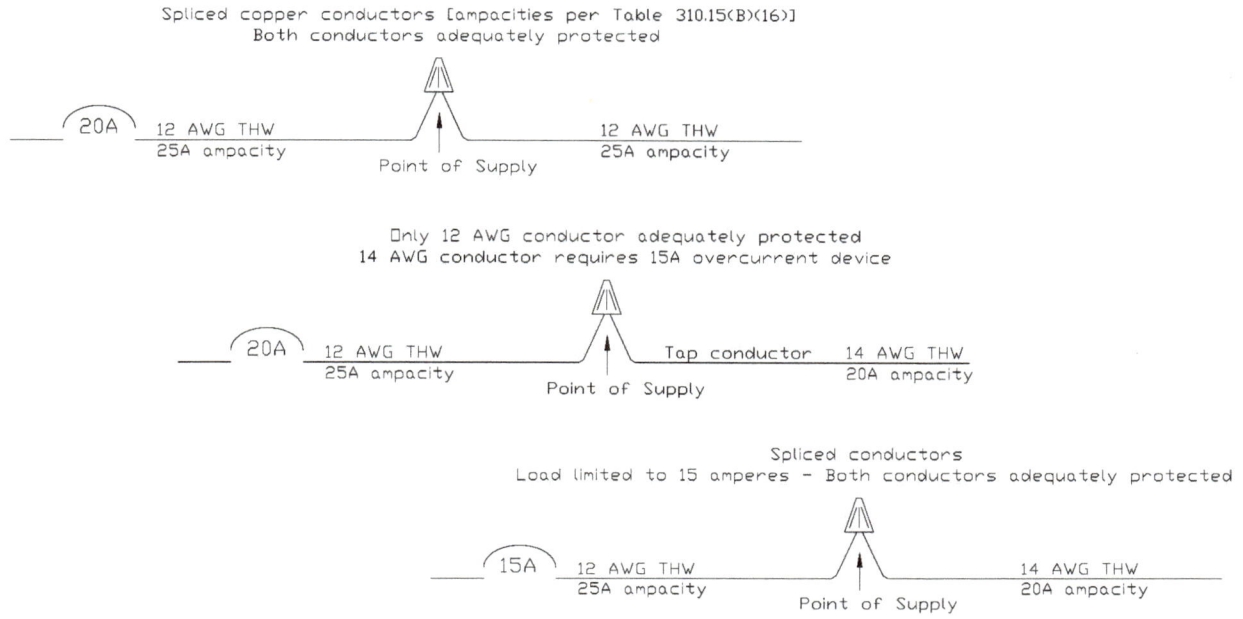

Spliced copper conductors [ampacities per Table 310.15(B)(16)]
Both conductors adequately protected

20A   12 AWG THW   12 AWG THW
       25A ampacity   25A ampacity
              Point of Supply

Only 12 AWG conductor adequately protected
14 AWG conductor requires 15A overcurrent device

20A   12 AWG THW   Tap conductor   14 AWG THW
       25A ampacity   20A ampacity
              Point of Supply

Spliced conductors
Load limited to 15 amperes - Both conductors adequately protected

15A   12 AWG THW   14 AWG THW
       25A ampacity   20A ampacity
              Point of Supply

Figure 240.2 - Spliced conductors/Tap conductors (at 75°C)

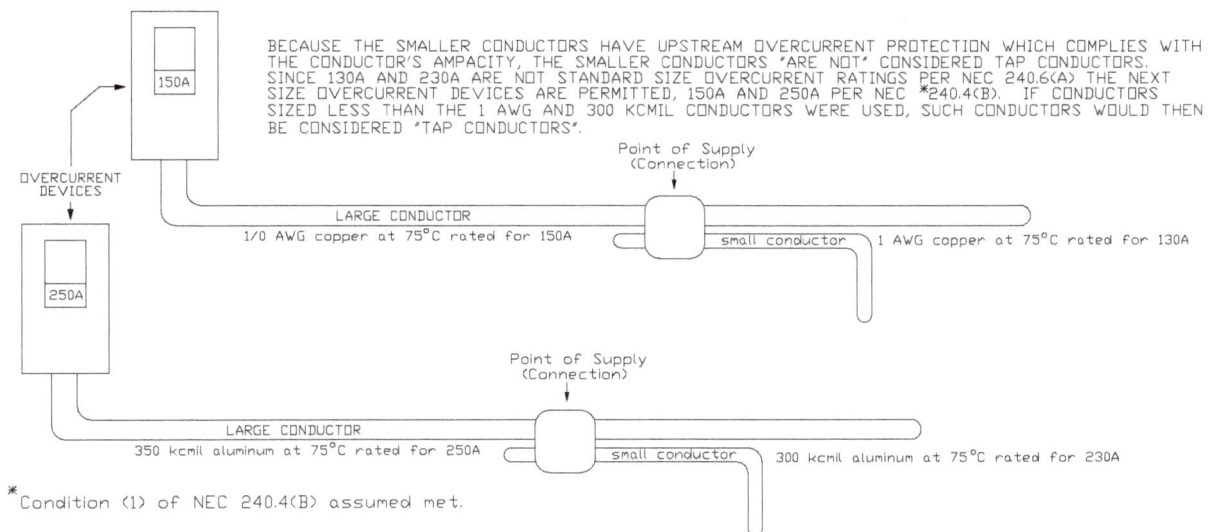

150A

OVERCURRENT
DEVICES

250A

BECAUSE THE SMALLER CONDUCTORS HAVE UPSTREAM OVERCURRENT PROTECTION WHICH COMPLIES WITH THE CONDUCTOR'S AMPACITY, THE SMALLER CONDUCTORS 'ARE NOT' CONSIDERED TAP CONDUCTORS. SINCE 130A AND 230A ARE NOT STANDARD SIZE OVERCURRENT RATINGS PER NEC 240.6(A) THE NEXT SIZE OVERCURRENT DEVICES ARE PERMITTED, 150A AND 250A PER NEC *240.4(B). IF CONDUCTORS SIZED LESS THAN THE 1 AWG AND 300 KCMIL CONDUCTORS WERE USED, SUCH CONDUCTORS WOULD THEN BE CONSIDERED 'TAP CONDUCTORS'.

Point of Supply
(Connection)

LARGE CONDUCTOR
1/0 AWG copper at 75°C rated for 150A   small conductor 1 AWG copper at 75°C rated for 130A

Point of Supply
(Connection)

LARGE CONDUCTOR
350 kcmil aluminum at 75°C rated for 250A   small conductor 300 kcmil aluminum at 75°C rated for 230A

*Condition (1) of NEC 240.4(B) assumed met.

Figure 240.3 - Smaller conductors *not always tap* conductors

According to NEC 240.21(B), where the feeder tap rules of 240.21(B)(1) through 240.21(B)(5) are followed, an overcurrent device is not needed at the point of supply (tap conductor to feeder conductor). Conductors permitted to be tapped without overcurrent protection *are not* permitted to be covered under the provisions of NEC 240.4(B). Because all feeder tap rules under the provisions of NEC 240.21(B), in one way or another require tap conductors to terminate in an overcurrent device - to limit the load to the ampacity of the tap conductors, *implies* the exclusion of NEC 240.4(B).

In most situations it's not always obvious as to why NEC 240.21(B) is titled "Feeder Taps" opposed to "Taps". By definition (Article 100), all circuit conductors between the service equipment, the source of a separately derived system, or other power supply source and the final branch-circuit overcurrent device, defines a *feeder*.

In reference to the definition, Figures 240.2 and .3 clearly illustrates an overcurrent device (other power supply) ahead of the tap conductors that primarily protects the conductors before the tap conductors (identifies all circuit conductors). Simply said, the supplying conductor "feed" the "tap" conductors. Therefore, to be in compliance with the definition of a *feeder*, tap conductors are required to terminate in an overcurrent device which could be the final overcurrent device that supplies a single branch circuit or one that leads to a final branch-circuit overcurrent device. For such reasons, NEC 240.21(B) is titled, "Feeder Taps". Although, Figure 240.2 was used as part of this discussion, it is a representation of a branch circuit where the tap or splice conductors could possibly supply an outlet. As for Figure 240.3, the smaller conductor which could also represent a "tap conductor" will terminate on an overcurrent protective device or a device, such as a main lugs only panelboard. To review illustrations of a branch and feeder circuit refer to Figures 210.1 and 215.1.

The phase, "to limit the load to the ampacity of the tap conductors", parallels to NEC 215.2(A)(1) where "feeder conductors shall have an ampacity not less than required to supply the load as calculated in Parts III, IV, and V of Article 220 and NEC 215.3 which requires the rating of a feeder overcurrent device to be no less than the loads being supplied.

With all being said, when applying the tap rules one must always remember that the use of the next higher standard overcurrent device that's rated above the ampacity of the tap conductors being protection is not permitted. Now let's proceed.

240.21(B)(1) - Taps Not Over 3 m (10 ft) Long (Feeder Taps) (13. - 15.) [3]

13. Refer to Figure 240.21(B)(1)-13. A three-phase feeder that's rated for 75°C is installed in rigid steel conduit (RSC) and protected by 300A fuses. A set of 7' tap conductors also enclosed in rigid steel conduit terminates on a 60A overcurrent device which protects a 52 amps load. Determine the size conductors needed to serve as the tap conductors.

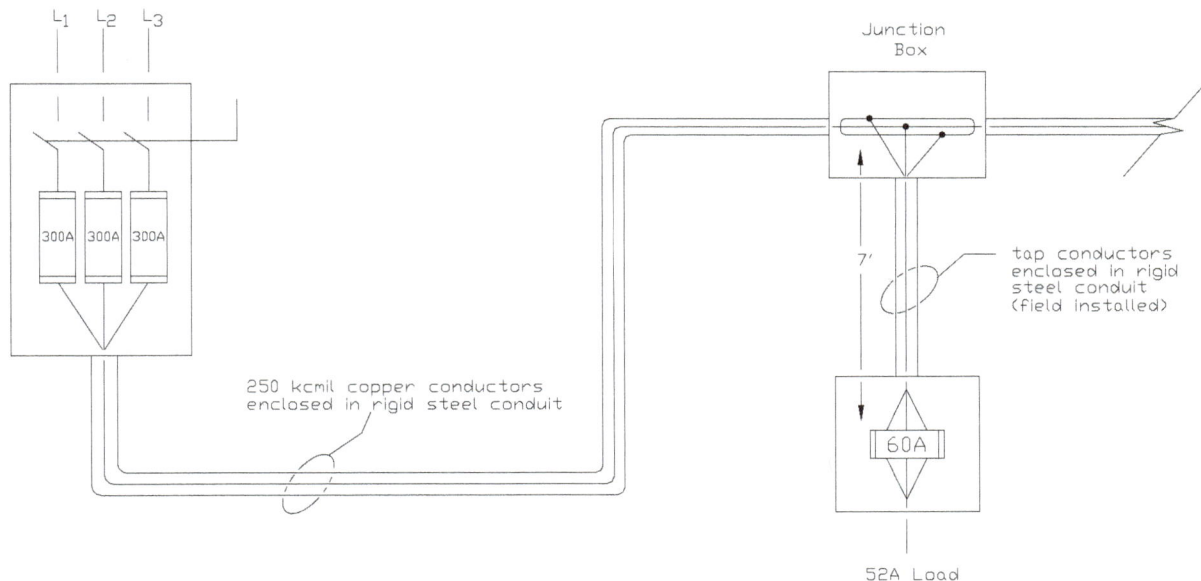

Figure 240.21(B)(1)-13 - Tap Not over 10 ft.

Resolving this problem step by step, let's began with condition **(1) a.** Condition **(1) a.** states that the ampacity of the tap conductors must not be less than the combined calculated loads on the circuits supplied by the tap conductors.

As a result, the tap conductors must be sized to supply the 52 amps load.

Because the tap conductors are being supplied by copper feeder conductors that's rated for 75°C per Table 310.15(B)(16), 6 AWG copper conductors with the same insulation type is also selected in addition to meeting the requirements of conditions **(1) a.** and **b.** of NEC 240.21(B)(1). The 6 AWG conductors have the ampacity (65A) to suffice the 52 amps load and are rated above the 60A overcurrent device they terminate in. Condition **(2)** is not applicable and the enclosed raceway requirements of condition **(3)** are met. As for condition **(4)**, being that this is a field installation per Figure 240.21(B)(1)-13, the ampacity of the tap conductors cannot be less than one-tenth of the rating of the overcurrent device protecting the feeder conductors, observe

$$300A \times 1/10 \text{ } or \text{ } 300A \times .10 \text{ } or \text{ } 300A/10 = 30A$$

At 65A, the ampacity of the 6 AWG tap conductors clearly exceeds one-tenth 1/10 (.10 or 10 percent) of the rating of the 300A overcurrent device protecting the feeder conductor where,

$$65A / 30A = 2.17 \text{ (times above 1/10)}$$

or

65A/300A = .217 or 21.7 percent (exceeds 10 percent)

which reflects the ampacity of the tap conductors being far greater than one-tenth of the 300A overcurrent device or 21.7 percent above the rating of the overcurrent device.

In such application(s) always keep in mind that the tap conductors are supplying the overcurrent device, meaning current will flow through the conductors until such current exceeds the rating of the supplied overcurrent device. Therefore, to further protect the tap conductors, the ampacity of the tap conductors must either equal or exceed the rating of the overcurrent device being supplied (terminated upon). Lastly, another thing to keep in mind is, tap conductors are always inadequately protected at its point of supply and the only means of protection in such case is through the terminating overcurrent device.

14. Refer to Figure 240.21(B)(1)-14. The referenced feeder and tap conductors are enclosed in conduit. All tap conductors are less than 10 feet. What size tap conductors are required?

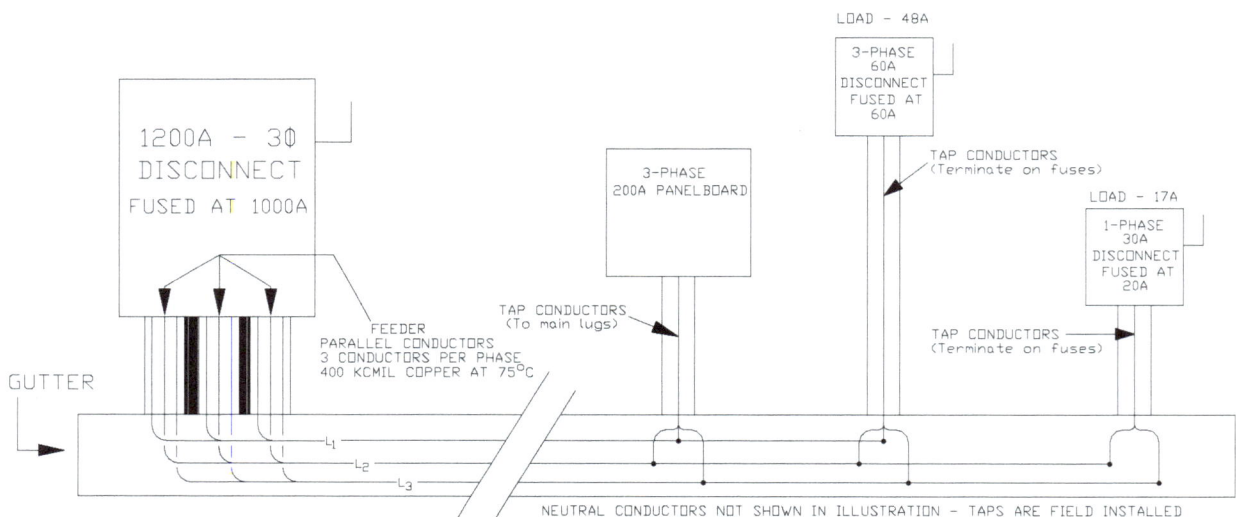

Figure 240.21(B)(1)-14 - Tap Not over 10 ft.

Considering condition **(1) a.** - Ampacity of tap conductors not less than load *and*
(1) b. - Ampacity of tap conductors not less than rating of device *or* not less than rating of overcurrent device.

To meet both requirements of condition **(1)** while applying similar type conductors (copper at 75°C), as a minimum, the following tap conductors could be used:

200A panelboard the device supplied by tap conductors - 3/0 AWG conductors rated for 200A
60A disconnect fused at 60A supplying a 48A load - 6 AWG conductors rated for 65A
30A disconnect fused at 20A supplying a 17A load - 12 AWG conductors rated for 25A
incompliance with NEC 240.4(D)(5)
[minimum]

— 32 —

Considering condition **(2)** - In compliance.
Considering condition **(3)** - In compliance, tap conductors are enclosed in raceway that extends from the tap to enclosures (panelboard, disconnects).

Considering condition **(4)** - Ampacity of tap conductors not less than one-tenth (1/10) of rating of feeder overcurrent device if field installed.

One-tenth of the 1000A feeder overcurrent device (fuses) results to,

$$1000A \times 1/10 \text{ or } 1000A/10 = 100A$$

Because the tap conductors are field installed and leave the enclosure of the gutter after taps were made, the ampacity of the tap conductors must be determined in accordance with condition **(4)** opposed to those sized per condition **(1)**. Based on the calculated results (as a minimum) the ampacity of each set of tap conductors cannot be rated less than 100 amps.

Therefore, the 3/0 AWG tap conductors supplying the 200A panelboard are permitted because they exceed the 100 amps requirement. As for the tap conductors' supplying the 60A and 30A disconnects, neither conductor size can be used. As a minimum 3 AWG conductors rated for 100 amps at 75°C must be used.

With the exception of the 200A panelboard, one thing to re-consider as to the requirement of condition **(4)** and that is, the ampacity of the tap conductors is the sole intend of this requirement, *no less than one-tenth of the rating of the overcurrent device protecting the feeder*. This requirement has no bearing upon the rating of an overcurrent device at the termination of the tap conductors, where in this case the 60A and 20A fuses. The rating of the terminating overcurrent device (or a device being supplied by tap conductors) *can be* less than one-tenth of the rating of the overcurrent device protecting the feeder. For an installation such as this, NEC 110.3(B) and 110.14(C) must also be adhered to.

15. A 9' tap is made from 208/120V, 3φ feeder conductors to supply a 225A, 3φ power panelboard (MLO) rated for 75°C. The ungrounded feeder conductors are 800 kcmil THW copper conductors and are protected by a 500A circuit breaker. Both feeder and tap conductors are enclosed in intermediate metal conduit. Determine the minimum size THW copper conductors per Table 310.15(B)(16) required to serve as the tap conductors.

Four-aught (4/0) AWG copper conductors rated for 75°C can be used as the required tap conductors. The conductors have an ampacity (230A) that exceeds the rating of the 225A panelboard (device) and all possible load demands. The tap conductors are enclosed in conduit and their ampacity is not less than one-tenth of the rating of the 500A feeder circuit breaker (500A/10 = 50A).

240.21(B)(2) - Taps Not Over 7.5 m (25 ft) Long (Feeder Taps) (16. - 17.) [2]

Unlike the 10 ft. rule, the ampacity of tap conductors over 10ft. and not more than 25 ft. is required to be one-third (1/3) the rating of the feeder overcurrent device opposed to one-tenth (1/10). One-third of the rating will required a conductor with a greater ampacity rating.

16. Refer to Figure 240.21(B)(2)-16. A three-phase feeder is protected by a 600 amps overcurrent device. The feeder is installed in a metal wireway. From the wireway the feeder supplies a run of tap conductors that extends 21' from their point of supply and terminates in a 150A fusible disconnect switch (FDS). The tap conductors are enclosed in electrical metal tubing. Determine the minimum size tap conductors needed for this installation. Also, determine the minimum size feeder conductors needed to supply the MLO panelboard. Consider the use of THWN copper conductors for both installations.

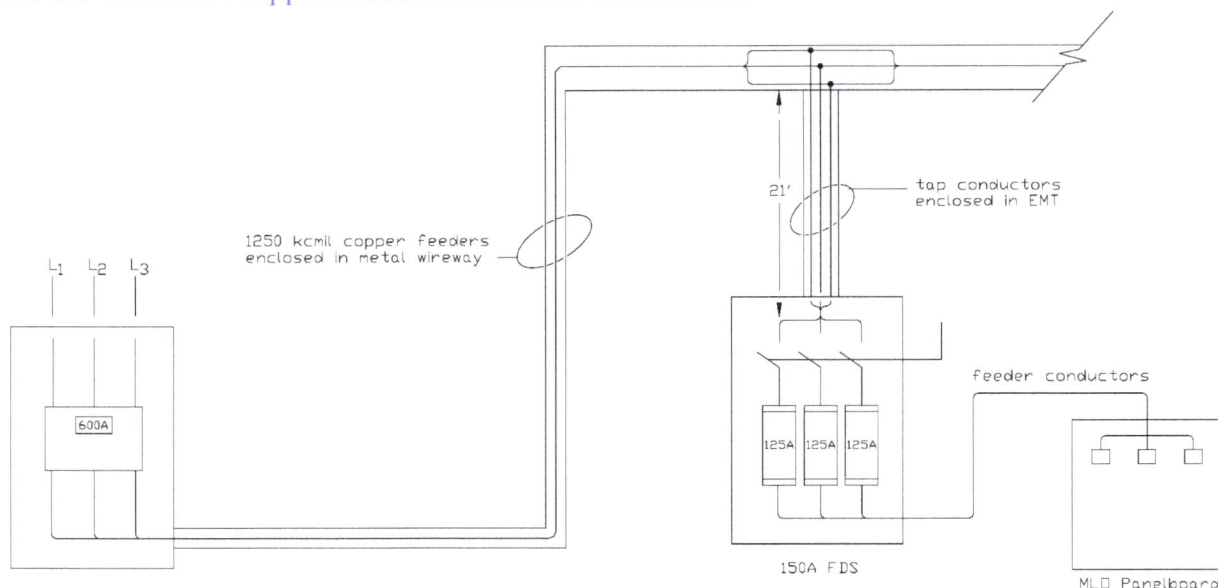

Figure 240.21(B)(2)-16 - Tap not over 25 ft.

In observing Figure 240.21(B)(2)-16 it should be clear that conditions **(2)** and **(3)** of NEC 240.21(B)(2) are met. Therefore, the only thing left is to determine the remaining requirement of condition **(1)**. Condition **(1)** requires the tap conductors needed for this installation to have an ampacity no less than one-third (1/3) the rating of the overcurrent device (600A fuses) protecting the feeder conductors. To determine such requirement the following calculation is performed,

$$600A \times 1/3 \text{ or } 600A/3 = 200A$$

based on the calculated results, 3/0 AWG THWN copper conductors which have a rated ampacity of 200A per Table 310.15(B)(16) are the minimum size tap conductors required for this installation.

As for the feeder conductors supplying the *main lugs only* panelboard, as a minimum 2 AWG THWN conductors (rated for 115A) are permitted per NEC 240.4(B). If the 150A disconnect switch was fused at 150A instead, 1/0 AWG THWN feeder conductors (rated for 150A) would be required based on the condition of NEC 240.4(B)(2). Assuming condition **(1)** of NEC 240.4(B) is met, the rated ampacity of the feeder conductors are permitted to be protected by the 125A fuses because the conductors are not tap conductors.

17. A 350 amps circuit breaker is used to protect 600 kcmil THW aluminum feeders. A set of conductors are tapped from the feeder conductors approximately 13.5' to supply a 100A panelboard with a main breaker. As a minimum, what size THW aluminum conductors are required to serve as the tap conductors? (The tap conductors are adequately protected from physical damage).

$$350 \text{ amps x } 1/3 = 116.7 \text{ amps}$$

Although the panelboard is only rated for 100A the tap conductors supplying the panelboard must have an ampacity rated no less than 116.7 amps. As a result, the minimum size conductors required for this application must be 1/0 AWG THW aluminum conductors (120A rated ampacity). Again, just as explained with the 10 ft. rule, a terminating overcurrent device (or the device being supplied by the tap conductors) as it pertains to the 25ft. rule *can be* rated less than one-third the rating of the feeder's overcurrent device.

240.21(B)(3) - Taps Supplying a Transformer [Primary Plus Secondary Not over 7.5 m (25 ft) Long] (Feeder Taps)

See discussion, **Transformer Secondary Tap Conductors** before proceeding [discussion follows 240.21(B)(5)]

18. Refer to the Figure 240.21(B)(3)-18A. The transformer's primary and secondary conductor's totals 23 feet in length. Determine the minimum size copper conductors (at 75°C) required for both primary and secondary tap conductors.

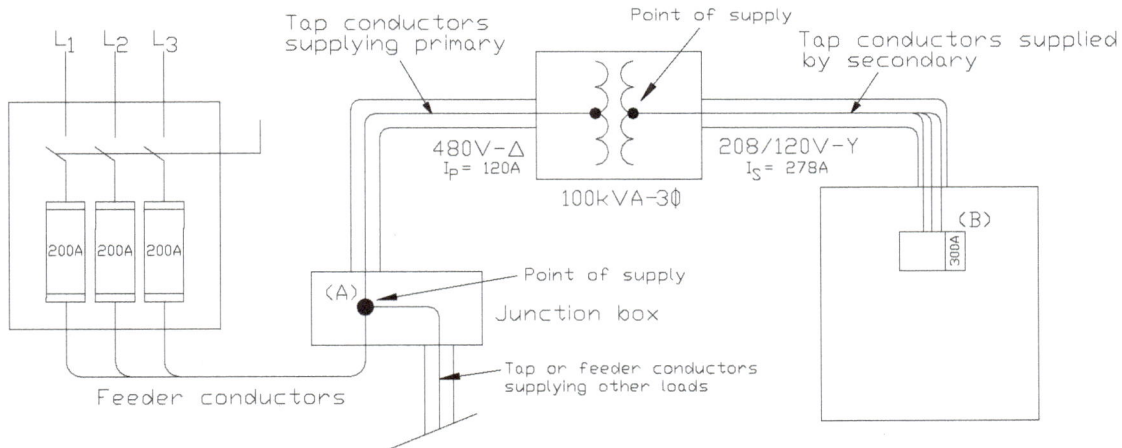

Total length primary (A) and secondary (B) conductors = 23 ft. Taps conductors are enclosed in approved raceway.

Figure 240.21(B)(3)-18A

Ensure conditions **(1)** and **(2)** of NEC 240.21(B)(3) are met:

Condition **(1)** - Primary tap conductors must have an ampacity at least one-third the rating of the overcurrent device protecting feeder conductors.

$$200A \times 1/3 = 66.7A$$

Per Table 310.15(B)(16), as a minimum the use of 4 AWG copper conductors rated for 85 amps at 75°C are required to meet condition **(1)**. However, the use of 4 AWG copper conductors are inadequate because they are incapable of carrying the transformer's primary full-load current at 120A. As a result, 1 AWG copper conductors which are rated for 130A at 75°C per Table 310.15(B)(16) are required. The rated ampacity of these conductors greatly exceeds the minimum requirements of condition **(1)**. Now, if the transformer supplied continuous loads where,

$$120A \times 1.25 = 150A$$

the use of 1/0 AWG copper conductors would then be required which have a rated ampacity of 150A at 75°C. Regardless of application, either set of conductors would still be considered tap conductors per definition.

Condition **(2)** - Secondary tap conductors must have an ampacity that is not less than (equal to or greater than) the value of the primary to secondary voltage ratio multiplied by one-third (1/3) of the rating of the overcurrent device protecting the feeder conductors.

As earlier determined, we know that one-third of the feeder's overcurrent protection is 66.7A.

The primary to secondary voltage ratio is,

$$480V / 208V = 2.31$$

As a result, the required ampacity of the secondary conductors can be determined where,

$$66.7A \times 2.31 = 154.08A$$

Per calculation, the ampacity of the secondary conductors must be either equal to or greater than 154.08A to satisfy the requirements of condition **(2)**. Therefore, per Table 310.15(B)(16), as a minimum, 2/0 AWG copper conductors rated for 175 amps at 75°C are required. However, the use of 2/0 AWG copper conductors would also prove be incapable of carrying the transformer's secondary full-load current at 278A. As a result, 300 kcmil copper conductors which are rated for 285A at 75°C per Table 310.15(B)(16) are required. In comparison to condition **(1)**, the rated ampacity of these conductors also greatly exceeds the minimum requirements of condition **(2)**. Again, if the transformer supplied continuous loads where,

$$278A \times 1.25 = 348A \text{ (rounded-up)}$$

the use of 500 kcmil copper conductors would then be required. At 75°C, the rated ampacity of these conductors is 380A.

Conditions **(3)** and **(4)** are satisfied per Figure 240.21(B)(3)-18A.

Condition **(5)** - Secondary tap conductors must terminate in a single circuit breaker or set of fuses that limit the load current to not more than the conductor ampacity that is permitted by 310.15.

Based upon the requirement of condition **(5)**, the ampacity of the secondary tap conductors must be either equal to or exceed 300A, the rating of the terminating overcurrent protective device. Earlier, it was determined per condition **(2)** that the minimum size secondary tap conductors required were 2/0 AWG copper; yet, this condition was further satisfied by the consideration of 500 kcmil copper conductors (380A) providing, the transformer supplied *continuous* loads. On the other hand, if the transformer only supplied *non-continuous* loads, 350 kcmil copper conductors which are rated for 310A must now be used to meet condition **(5)** opposed to the required use of 300 kcmil copper conductors (285A). Although the 300 kcmil conductors had an ampacity to suffice the non-continuous load (278A), the provisions of NEC 240.21(C) does not permit such application [ampacity {285A} less than rating of overcurrent protection {300A} – as permitted in NEC 240.4(B)].

Now, let's take a look at one other thing. Until now the size of the feeder conductors has never been discussed however, there are a set of conductors that *does* correspond with the ampere rating of the feeder's overcurrent device which would counter NEC 240.4(B)(2). At 75°C per Table 310.15(B)(16), 3/0 AWG copper conductors are also rated for 200A. Refer now to Figure 240.21(B)(3)-18B.

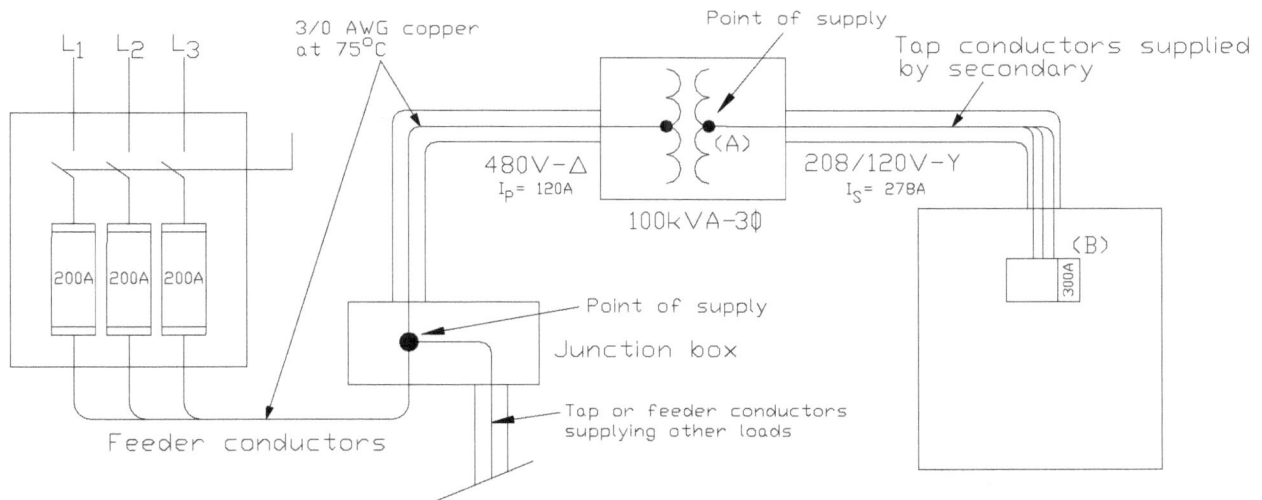

Total length of secondary conductors can now extend the entire 23 ft. but not exceed 25 ft. Tap conductors are enclosed in approved raceway.

Figure 240.21(B)(3)-18B

Figure 240.21(B)(3)-18B reflects the once smaller size tap conductors being sized at 3/0 AWG copper rated for 75°C. Since the conductors supplying the transformer's primary are now the same size as the feeder conductors and share likewise overcurrent protection, the provision of condition **(1)** is no longer a factor as it once was. Because the primary conductors are now protected at their rated ampacity, according to condition **(3)** the transformer's secondary conductors can now extend the entire 23 ft. length because the length of the primary conductors can now be excluded from the total length.

In concluding, where the latter portion of condition **(5)** states, *"that is permitted by 310.15"* this simply means all applicable conductor ampacity tables as listed according to NEC 310.15(B) or per NEC 310.15(C), the calculated conductor ampacities under engineering supervision, to include correction and adjustment factors as needed to produce a conductor ampacity that will safely supply the load while operating within the conductor's permitted temperature rating. Where applicable, other factors such as device termination ratings along with continuous or noncontinuous loads or both must also be considered.

240.21(B)(4) - Taps Over 7.5m (25 ft) Long (Feeder Taps)

When an electrical installation requires feeder tap conductors to extend over 25 ft. but not more than 100 ft., this tap rule is the one to use. In addition to the increased length such application relates to **(a)** high bay manufacturing buildings with walls over 35 ft. where the building's electrical system is **(b)** maintained and supervised by qualified personnel. Other related conditions requires the feeder tap conductors to **(c)** be run no more than 25 feet horizontally (including bends on the same horizontal plane), **(d)** have an ampacity no less than one-third (1/3) the rating of the feeder overcurrent device, **(e)** terminate in a single circuit breaker or set of fuses, **(f)** be enclosed in approved raceway, **(g)** be continuous without splices, **(h)** be no smaller than 6 AWG copper or 4 AWG aluminum, **(i)** not penetrate walls, floors or ceilings and **(j)** be made no less than 30 ft. from the floor.

19. A 2/0 AWG THWN copper feeder enclosed in rigid metal conduit is protected by a 150A, 3φ circuit breaker that is fed from a 400A switchboard. The feeder is located 48 feet above the floor in an industrial building and terminates in a metal junction box where tap conductors are supplied by the feeder conductors. Enclosed in electrical metallic tubing, the tap conductors are ran 16.5 feet vertically beneath the tap conductors' point of supply and terminate in a 100A, 3φ panelboard that's located 21 feet to the right of the junction box. The panelboard is protected by an 80A main circuit breaker. Both feeder and tap conductors will be type THWN. The terminal rating of the circuit breakers and switchboard is 75°C. Determine the minimum size tap conductors needed for this installation.

Although an illustration is not provided with this question, the best approach would be to sketch the installation as explained to ensure all 9 conditions of NEC 240.21(B)(4) are met. However, another means is to analyze each condition individually, which is the approach that's taken here. To begin,

Condition **(1)** - The assumption is that this condition is met.

Condition **(2)** - Horizontally, the tap conductors extends 21 ft. which is less than permitted. Combined with the 16.5 ft. vertical distance totals 37.5' in length which is less than the 100 ft. limitation.

Condition **(3)** - To determine the minimum size tap conductors based upon the 150A overcurrent device protecting the feeder where,

$$150A \times 1/3 = 50A$$

Based upon the calculation, as a minimum, 8 AWG copper or 6 AWG aluminum THWN conductors rated for 75°C are required to suffice condition **(3)**. Per Table 310.15(B)(16), the ampacity of both conductors is 50A.

Condition **(4)** - The tap conductors terminate in an 80A main circuit breaker of the 100A, 3φ panelboard. Because Condition **(4)** requires the terminating overcurrent device to *"limit the load to the ampacity of the tap conductors"*, the 8 AWG copper or 6 AWG aluminum conductors derived in Condition **(3)** cannot be used under the current condition. To adhere to Condition **(4)** two things can be done, they are: **(a)** Upgrade the existing main circuit breaker to 100A to maximize the use of the 100A panelboard resulting to the use of 3 AWG copper or 1 AWG aluminum THWN conductors at 75°C where both conductors are rated for 100A *or* **(b)** As a minimum, use 4 AWG copper (85A) or 2 AWG aluminum (90A) THWN conductors to use the existing 80A main circuit breaker. Again, as a reminder, the provisions of NEC 240.4(B) are not permitted.

Condition **(5)** - The tap conductors are enclosed in electrical metallic tubing (EMT).

Condition **(6)** - The tap conductors run continuous from their point of supply to the 80A main circuit breaker of the 100A panelboard, no splices involved.

Condition **(7)** - Regardless of the resulting calculation or load as it pertains to Condition **(3)**, when using this tap rule the smallest tap conductors that can be used for such installation must be 6 AWG copper or 4 AWG aluminum. There is no compromise.

Condition **(8)** - The tap conductors do not penetrate walls, floors, or ceilings. However, this restriction does not apply to feeder conductors.

Condition **(9)** - The tap is made 31.5 ft. (48 ft. − 16.5 ft.) above the floor exceeding the 30 ft. minimum.

240.21(B)(5) - Outside Taps of Unlimited Length (Feeder Taps)

Where feeder tap conductors or transformer secondary conductors are located on the outside of a building or structure, the length of the conductors is permitted to be unlimited.

20. Although there are no required calculations for this tap rule, the four (4) listed conditions of NEC 240.21(B)(5) are illustrated in Figure 240.21(B)(5)-20A and B for clarity.

Condition **(1) [C1]** - Requires the feeder tap or transformer secondary conductors to be protected from physical damage whether installed above or underground.

Condition **(2) [C2]** - Requires the feeder tap or transformer secondary conductors to be terminated in a single circuit breaker or set of fuses that limit the load to the ampacity of the tap conductors, meaning "the ampacity of the tap conductors must be equal to or greater than the rating of the terminating overcurrent device". The single overcurrent device is permitted to supply any number of additional overcurrent devices on its load side. The requirements of Condition **(2)** are identical to the conditions found in NEC 240.21(B)(2)(2) and 240.21(B)(4)(4).

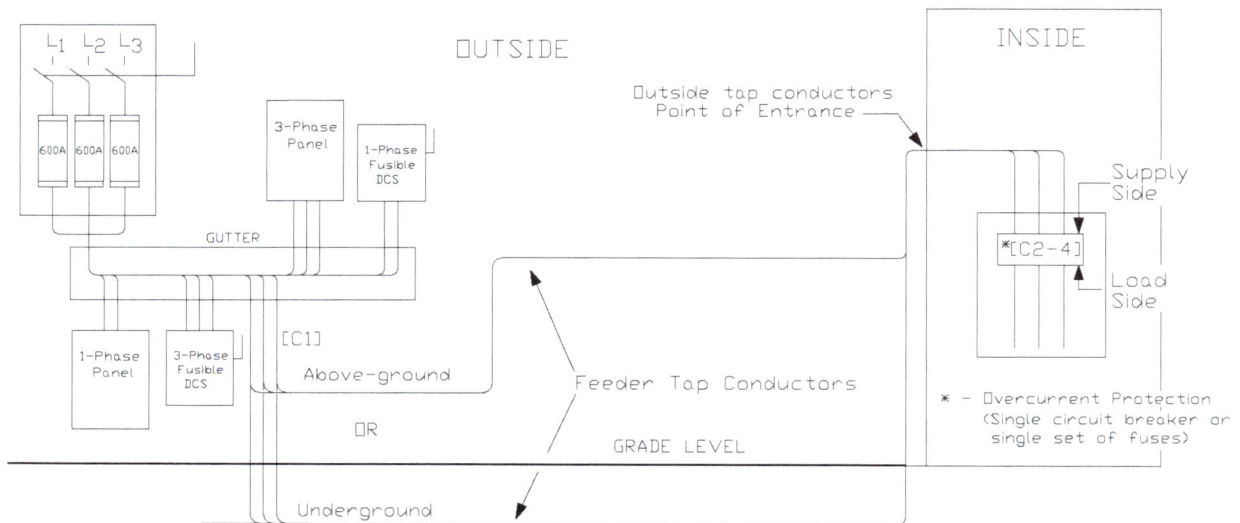

Figure 240.21(B)(5)-20A - Outside Feeder Taps of Unlimited Length

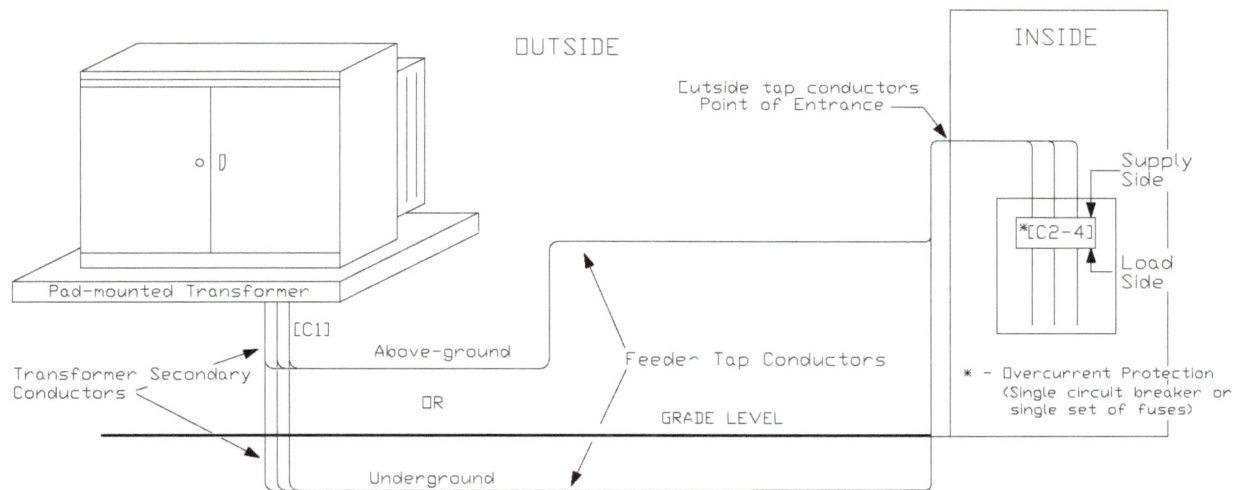

Figure 240.21(B)(5)-20B - Outside Transformer Secondary Conductors of Unlimited Length

Condition **(3) [C3]** - Requires the overcurrent device protecting the feeder tap or transformer secondary conductors to be installed within a disconnecting means (panelboard with main circuit breaker, circuit breaker disconnect switch or a fusible disconnect switch) or located near (in close proximity to) the disconnecting means.

Condition **(4) [C4]** - Requires the feeder tap or transformer secondary conductors to be installed outside (whether above or underground) and terminated likewise. Feeder tap or transformer secondary tap conductors are only permitted indoors at the point where they enter a building (point of entrance). Disconnecting means for feeder tap or transformer secondary conductors should be as close as possible to point of entrance. Disconnecting means for the conductors must be installed at a readily accessible location.

Transformer Secondary Tap Conductors

Similar to the provisions of NEC 240.21(B) where the primary focus is *feeder tap conductors*, NEC 240.21(C) provides exclusive provisions for *transformer secondary conductors*. NEC 240.21(C) permits a set of conductors feeding a single load, or each set of conductors feeding separate loads to be connected to a transformer secondary without overcurrent protection at the secondary, refer to Figures 240.4(a) and (b).

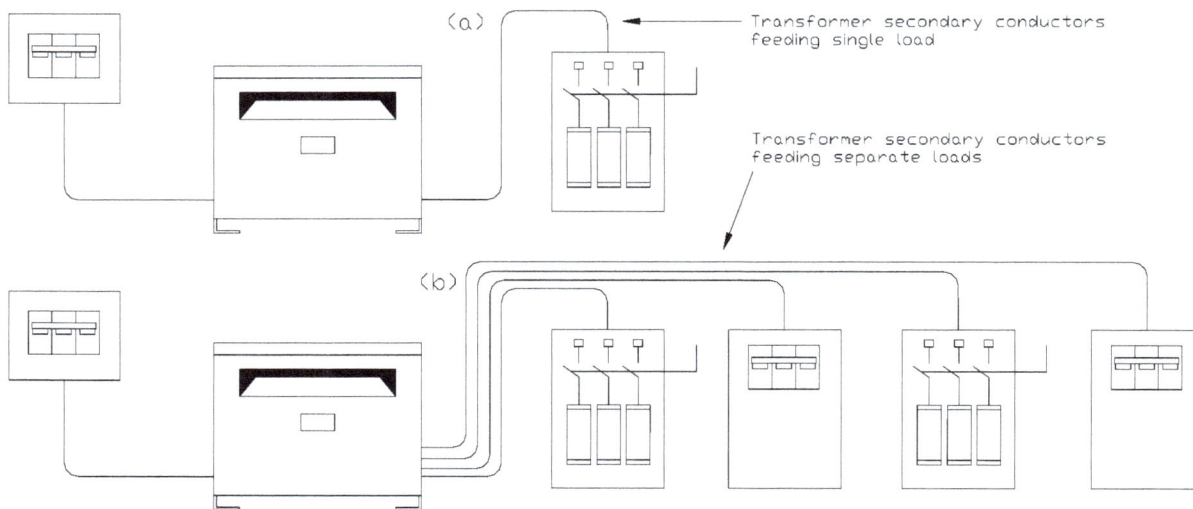

Figures 240.4(a) and (b)

When secondary conductors are supplied from the terminals of a transformer's secondary they are unprotected and identified by definition as "tap conductors", observe Figure 240.5(a).

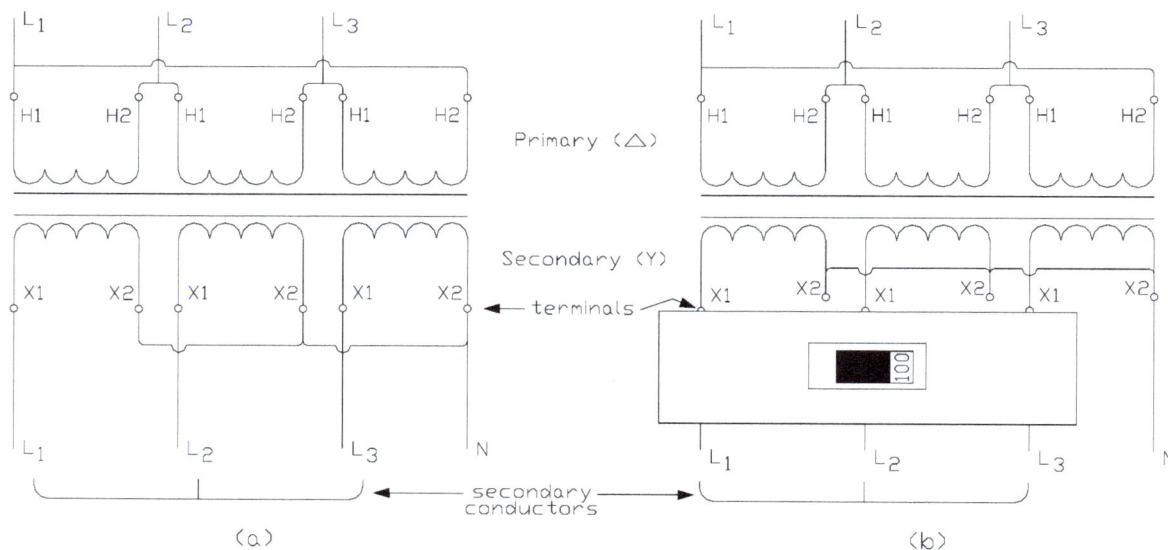

240.5 - (a) Tap / (b) Protected - Secondary conductors

If an overcurrent device could be installed at the secondary terminals of a transformer, thus eliminating a direct connection to the terminals, transformer secondary conductors would no longer be classified as "tap conductors". Hypothetically, Figure 240.5(b) displays the transformer's secondary conductors originating from an overcurrent device and therefore protected. Now, if the secondary conductors were to terminate in a main lugs only (MLO) circuit breaker panelboard for example, the conductors by definition would be recognized as "feeder conductors" unlike those shown in Figure 240.5(a).

NEC sub-sections 240.21(C)(1) through (C)(6) provides specific guidelines to ensure compliance. Where transformer secondary conductors are permitted to be tapped without overcurrent protection, the provisions of NEC 240.4(B) are not applicable. Just as NEC 240.21(C) provides specific instructions for *transformer secondary conductors*, NEC 450.3 does likewise for specifying *overcurrent protection requirements for transformers*. Just as a set of transformer secondary conductors are permitted to feed separate loads, Tables 450.3(A) and (B), **Note 2** permits and specify a maximum number of secondary overcurrent devices that one transformer can feed. According to **Note 2**, where secondary overcurrent protection is required, the secondary overcurrent device shall be permitted to consist of not more than six circuit breakers or six sets of fuses grouped in one location. The six secondary overcurrent devices permitted in **Note 2** take on a similar allowance to that of NEC 230.71(A) where a maximum of six service disconnects are permitted be grouped in any one location per service. In Figure 240.4(b), two additional secondary overcurrent devices are permitted.

For questions pertaining to Tables 450.3(A) and (B), **Note 2** see question Nos. 9. and 10. of Article 450 (Volume 3).

Although secondary conductors may terminate in a secondary overcurrent device as shown in Figures 240.4(a), (b) and 240.5(a), the conductors aren't actually protected because their only protection is totally dependent upon the activation of an upstream primary overcurrent device. In order for secondary conductors to be totally protected, the conductors would have to originate from an overcurrent device, that is, be directly attached to the secondary terminals of a transformer as shown in Figure 240.5(b). If a short circuit or ground fault condition occurred between the transformer's terminals (secondary conductor's point of supply) and secondary overcurrent device [Figures 240.4(a) and (b)], chances of the transformer's primary overcurrent protection device tripping to distinguish such conditions are uncertain. For this reason, the transformer tap rules in sub-section (C) of NEC 240.21 were developed to reduce the risk of such dangerous conditions. These rules limits the maximum distance between transformer secondary conductors and secondary overcurrent devices or a transformer's primary and secondary conductors combined.

NEC 240.21(C) only allows a transformer's secondary conductors to go unprotected based on six conditions,

(1) A transformer's primary can only protect the secondary conductors when they (secondary conductors) are supplied from a 2-wire, single-phase transformer (examples: door bell, irrigation control, landscape lighting, motor control transformers), or a 3-wire, three-phase delta-delta transformer (no neutral conductors) *or* **(2)** when a transformer's secondary conductors will not extend more than 10 feet from where they are connected to the transformer's secondary side *or* **(3)** for industrial installations when a transformer's secondary conductors will not extend more than 25 feet from where they are connected to the transformer's secondary side *or* **(4)** when a transformer's secondary conductors will be located outside *or* **(5)** when a transformer's secondary conductors originates from a feeder tapped transformer *or* lastly **(6)** for non-industrial installations when a transformer's secondary conductors will not extend more than 25 feet from where they are connected to the transformer's secondary side.

Although Tables 450.3(A) and (B) provides values for determining the maximum overcurrent devices for transformer (windings) protection, these devices may not necessarily provide adequate protection for conductors that either supply or receive current from a transformer.

240.21(C)(1) - Protection by Primary Overcurrent Device (Transformer Secondary Conductors)

21. A 100kVA, 3φ transformer is rated for 500V-250V (Δ-Δ/3W). The transformer's primary is protected by a 150A circuit breaker. Determine the minimum ampacity of the secondary conductors if the conductors are solely protected by the transformer's primary overcurrent device. Also, determine the secondary conductors' minimum ampacity if the transformer's rated voltage was 600V-230V instead.

To begin this process let's first determine the transformer's primary and secondary full-load currents.

$$I_P = \frac{100,000VA}{500V \times \sqrt{3}} = 115.5A \qquad I_S = \frac{100,000VA}{250V \times \sqrt{3}} = 231A$$

NEC 240.21(C)(1) allows the secondary conductors (3-wire, single voltage) when supplied from a delta-delta connected transformer to be protected by the transformer's primary overcurrent device if the device is sized according to NEC 450.3 [Table 450.3(B) – **Currents of 9 Amperes or More** – Primary only protection]. Because the transformer's primary overcurrent device (150A) is sized at 125 percent (1.25) of the transformer's primary full-load current (144.4) (115.5A x 1.25), the protection is in accordance with NEC 450.3.

NEC 240.21(C)(1) also requires the rating of the primary overcurrent device to be either less than or equal to the ampacity of the transformer's secondary conductors when multiplied by the secondary to primary transformer voltage ratio. To determine the ampacity of the secondary conductors to comply with this requirement, observe

$$250V \ / \ 500V \times \mathbf{A} = 150A \text{ (primary device)}$$

where **A** determines the required ampacity of the secondary conductors. Therefore,

$$\mathbf{A} = \frac{150A}{250V \ / \ 500V} = 300A$$

which means, the ampacity of the secondary conductors must be either equal to or greater than 300A in order to meet or exceed the requirements of NEC 240.21(C)(1). To verify the results, let's select the first copper conductor at 75°C that's listed in Table 310.15(B)(16) that is either less than or exceed 300A. Per Table 310.15(B)(16), a 300 kcmil and a 350 kcmil copper conductor are selected where the rated ampacities are 285A and 310A respectively. Applying the given formula for both where,

250V / 500V x 285A = 142.5A (150A primary overcurrent device exceed value [142.5A])
250V / 500V x 310A = 155A (150A primary overcurrent device does not exceed value [155A])

Based on the calculated results, if 300 kcmil copper conductors are used which represents an ampacity less than 300A, the 150A primary protection alone cannot serve as the sole means for the secondary conductors therefore, secondary protection is required. In contrast, if 350 kcmil copper conductors are used which represents an ampacity that exceeds 300A, the 150A primary protection can serve as the sole means for the secondary conductors therefore, secondary protection is not required.

In conclusion, if a 600V-230V voltage rating is applied where,

$$A = \frac{150A}{230V / 600V} = 391.3A$$

As a minimum, the ampacity of the secondary conductors must be either equal to or greater than 391.3A to again meet or exceed the requirements of NEC 240.21(C)(1).

240.21(C)(2) - Transformer Secondary Conductors Not Over 3 m (10 ft) Long (Transformer Secondary Conductors)

22. A 100kVA, 3φ transformer is rated for 480V-240/120V (Δ-Δ/4W). The transformer's primary is protected by a 150A circuit breaker. Secondary tap conductors enclosed in Type LFMC raceway are ran 9' to a 225A 3φ panelboard with a similar rated main breaker. The panelboard will be used to supply a 210A computed load. Determine the minimum ampacity of the required secondary (tap) conductors. Assume the use of copper conductors rated for 75°C.

Condition (1) b. of NEC 240.21(C)(2) is considered to determine the minimum ampacity of the secondary conductors. It states that the ampacity of the secondary conductors must not be less than the device (example: main lugs only panelboard) supplied by the secondary conductors or not less than the rating of the overcurrent-protective device at the termination of the secondary conductors. In this situation because the panelboard is equipped with a main breaker, as a minimum, 4/0 AWG secondary tap conductors rated for 230A, measuring 9' in length, must be used to terminate in the 225A main breaker. This installation will satisfy conditions (1) b. and (1) a., where the 4/0 AWG conductors is rated to accommodate the 210A computed load.

With the secondary conductors being enclosed in Type LFMC raceway and terminating in a panelboard, conditions (2) and (3) are also met.

Lastly, condition (4) which relates to field installations states that the 150A primary overcurrent device protecting the transformer's primary when multiplied by the primary to secondary transformer voltage ratio (480V/240V) must not exceed 10 times the ampacity of the secondary conductors. Therefore, considering the product (300A) of such requirements (150A x 480V/240V) and the ampacity of the secondary conductors (230A) where, 300A/230A = 1.3 reflects a value that does not exceed 10 times the ampacity of the secondary conductors when the rating of the overcurrent device protecting the transformer's primary is multiplied by the primary to secondary transformer voltage ratio.

240.21(C)(3) - Industrial Installation Secondary Conductors Not Over 7.5 m (25 ft) Long (Transformer Secondary Conductors)

23. A 1200A three-phase distribution panelboard is directly fed by a three-phase 1000kVA transformer in a steel manufacturing company. The transformer is rated for 2300-480/277V and like the company's entire electrical system is maintained by only qualified personnel. The panelboard is equipped with seven circuit breakers. The ratings of the circuit breakers are: 400A, 350A, 200A (2), 150A, and 125A (2). If the length of the transformer's secondary conductors extend 22 feet to the distribution panelboard, what size copper conductors at 75°C are required? Assume the secondary conductors will be enclosed in an approved raceway.

For this tap rule the only calculation involves that of Condition **(2)**. Because all overcurrent devices are installed in the distribution panelboard they are considered grouped thus meeting the requirements of Condition **(3)**. Now getting back to resolving Condition **(2)** where *first* the ampacity of the secondary conductors are required to be equal to or greater than the secondary current rating of the transformer. Per calculation, the transformer's secondary current rating yields,

$$I_S = \frac{1,000,000VA}{480V \times 1.732} = 1202.85A$$

Secondly, the sum of the ratings of the overcurrent devices does not exceed the ampacity of the secondary conductors where,

$$400A + 350A + 200A \times 2 + 150A + 125A \times 2 = 1550A$$

Now that the results of both requirements have been derived, the secondary conductors can be sized. Table 310.15(B)(16), list a 2000 kcmil copper conductor as having the largest ampacity of all 75°C copper conductors at 665A. Therefore, based on the derived results (1202.85A and 1550A), to satisfy either requirement the use of parallel conductors are required. Considering the results, a combination consisting of no less than three parallel conductors per line must be used. Although larger combinations can be used, the three conductor minimum will be applied. By dividing each result by three the required ampacity can be determined to determine the size secondary conductors needed. As a result,

$$\frac{1202.85A}{3} = 400.95 \qquad \frac{1550A}{3} = 516.67$$

At 400.95A, a 600 kcmil copper conductor at 420A is required. Not only does the ampacity of this conductor exceed 400.95A but when multiplied by three totals, 1260A (420A x3) which exceeds the transformer's secondary current rating, 1202.85A.

At 516.67A, a 900 kcmil copper conductor at 520A is required. Not only does the ampacity of this conductor exceed 516.67A but when multiplied by three totals, 1560A (520A x3) which exceeds the sum of the ratings of the overcurrent devices, 1550A.

To satisfy Condition **(2)** the 600 kcmil copper conductors cannot be used because the total ampacity of the three conductors (1260A) will only exceed the transformer's secondary current rating (1202.85A) while falling below the sum of the ratings of the overcurrent devices (1550A). In order to satisfy Condition **(2)**, three 900 kcmil copper conductors per line must be used.

Although the sum of the overcurrent devices exceeds the rating of the distribution panelboard, just remember the rating of a panelboard is based upon an assumed or calculated load and not the summation of selected overcurrent devices. Most overcurrent devices are determined based upon a percentage value that will exceed the load being protected.

240.21(C)(4) - Outside Secondary of Building or Structure Conductors (Transformer Secondary Conductors)

Just as there are no required calculations in NEC 240.21(B)(5) for "feeder tap" conductors the same conditions are applied when the "secondary conductors of a transformer" are located on the outside of a building or structure and ran at an unlimited length without means of overcurrent protection at the point where the conductors are connected to the transformer (point of supply). Refer to question No. 20. of section 240.21(B)(5) and Figure 240.21(B)(5)-20(B). The same four **(4)** conditions as listed in question No. 20. are applicable to this section, 240.21(C)(4).

240.21(C)(5) - Secondary Conductors from a Feeder Tapped Transformer (Transformer Secondary Conductors)

NEC 240.21(C)(5) reference the use of NEC 240.21(B)(3). Therefore, for those applications where tap conductors are used to supply a transformer the provisions of NEC 240.21(B)(3) can be applied. See question No. 18. of Section 240.21(B)(3).

240.21(C)(6) - Secondary Conductors Not Over (7.5 m) 25 ft Long (24. - 26.) [3]

24. A 200kVA, 3ϕ transformer is rated for 480V-208/120V (Δ-Y). The transformer's primary overcurrent protection is provided by a set of 250A time-delay fuses. Multiple sets of secondary tap conductors enclosed in metal raceways are run 17' to supply a 150A fusible disconnect switch fused at 125A and 22' to supply two panelboards rated for 200A and 225A. Both panelboards are equipped with main breakers. Determine the minimum ampacity of each set of secondary conductors.

Multiple sets of transformer secondary conductors supplying separate loads are permitted according to NEC 240.21(C). NEC 240.21(C)(6) only requirements are restricting the maximum length of the secondary conductors to 25 feet in addition to the following conditions,

(1) the secondary conductors must have an ampacity that is equal to or greater than the ratio of the primary to secondary voltage times one-third (1/3) the rating of the transformer's primary overcurrent device,

(2) the secondary conductors must terminate in a single circuit breaker or set of fuses that limits the load current to not more than the conductor ampacity permitted by NEC 310.15, *and*

(3) the secondary conductors must be suitably protected from physical damage.

Considering that the length of the secondary conductor's in this situation along with condition (3) are satisfied, let's focus on conditions (1) and (2).

Condition (1)

Let's determine the minimum ampacity of each set of secondary conductors by first considering the primary overcurrent protection at one-third of its rating.

$$250A \times 1/3 = 83.33A$$

Now when the minimum ampacity of each set of secondary conductors is multiplied by the primary to secondary voltage ratio, the results will not be less 83.33A. Once again, observe

$$480V / 208V \times 83.33A = \mathbf{A}$$

where **A** determines the required ampacity of the secondary conductors. Therefore,

$$\mathbf{A} = 2.31 \times 83.33A = 192.3 \text{ amps}$$

which means the minimum ampacity of each set of secondary conductors must be either equal to or greater than 192.3 amps to meet the requirements of condition (1).

Condition (2)

Because the secondary conductors will terminate in devices having overcurrent protection, the secondary conductors must have ampacities that will suffice the load current being protected by each device according to condition (2). Again, the required ampacity of each set of secondary conductors must not be less than 192.3A.

In this situation, the 200A and 225A overcurrent devices will require conductors with an ampacities that are either equal to or exceed their rating which would exceed the 192.3A minimal ampacity requirement per condition (2). As for the 150A disconnect switch, conductors with an ampacity that is either equal to or exceed 192.3A are required.

25. A 2-wire, 480/120V single-phase 1kVA control transformer is used to supply a motor control circuit. If the transformer's secondary conductors are ran 18' to supply the circuit, what size copper conductors are required?

The provisions of NEC 240.21(C)(1) permits such 2-wire transformer installations to be protected by the overcurrent protection provided on the primary side of the transformer per NEC 450.3. NEC 430.72(C)(4) reference the permitted use of a *maximum* percentage (500 percent) when a control circuit transformer primary current is rated *less than 2 amperes* opposed to Table 450.3(B) which list a *minimum* percentage (300 percent). Because the secondary conductors will be extended 18' beyond their point of supply, the conditions of NEC 240.21(C)(6) must be applied.

Before getting started let's size the transformer's primary overcurrent device per NEC 450.3. However, the transformer's primary full-load current must first be determined,

$$I_P = \frac{1000VA}{480V} = 2.08A$$

According to Table 450.3(B) when primary only protection is applied (as referenced in NEC 240.21(C)(1) for such transformer installation) a transformer's primary full-load current must be increased by 167 percent (1.67) where the primary current is 9 amps or less. Because the ampere rating exceeds 2 amperes, although slightly, NEC 430.72(C)(4) or Table 450.3(B) is not applicable. Therefore,

$$2.08A \times 1.67 = 3.47A$$

The primary overcurrent device is limited to 3A because there are no provisions to increase it to the next standard size. NEC 240.6(A) only list a 3 amps fuse as having a standard amperes rating. Now the applicable conditions of NEC 240.21(C)(6) can be applied.

Condition (1) - Again, let's determine the minimum ampacity of the secondary conductors by first considering the primary overcurrent protection at one-third of its rating.

$$3A \times 1/3 = 1A$$

Now when the minimum ampacity of the secondary conductors is multiplied by the primary to secondary voltage ratio, the results will not be less 4A. Once again, observe

$$480V / 120V \times 1A = \mathbf{A}$$

where, **A** determines the required ampacity of the secondary conductors. Therefore,

$$\mathbf{A} = 4 \times 1A = 4 \text{ amps}$$

which means, the minimum ampacity of the secondary conductors must be either equal to or greater than 4 amps to meet the requirements of condition (1). As a result, the minimum size conductors required must be 14 AWG. Considering all temperature ratings in Table 310.15(B)(16) for 14 AWG copper, all rated ampacities exceeds 4 amps.

Condition (2) - Not applicable. Per NEC 240.21(C), a set of conductors feeding a single load, is permitted to be connected to a transformer secondary, without overcurrent protection at the secondary, as specified in NEC 240.21(C)(6).

Condition (3) - Assumed that this installation will be in compliance with this condition.

26.   Refer to the transformer in question No. 28. of Article 430 (Volume 3).   From the transformer to the lighting panelboard, what size secondary copper conductors are required?

Both NEC 240.21(C)(3) and (C)(6) reference secondary conductors not over 25 feet in length. However, NEC 240.21(C)(6) will be referenced because the installation in this question is assumed to be non-industrial.

First acquiring the transformer's secondary full-load current,

$$I_S = \frac{15kVA\ (15{,}000VA)}{208V} = 72.12A$$

Now according to NEC 240.21(C)(6), condition **(1)** requires the secondary conductors to have an ampacity that is not less than the value of the primary-to-secondary voltage ratio multiplied by one-third (1/3) of the rating of the overcurrent device protecting the primary of the transformer.

To get started the primary-to-secondary voltage ratio is,

$$480V{:}208V \text{ or } \frac{480V}{208V} = 2.31$$

This ratio is multiplied by one-third of the rating of the overcurrent device protecting the transformer's primary as determined in question No. 8. of Article 450 (Volume 3).

Based on the primary (windings) being protected by a 40A overcurrent device the secondary conductor's ampacity can be no less than,

$$2.31 \text{ x } 40A \text{ x } 1/3 = 30.8A$$

Now let's consider the transformer's full-load current to size the secondary conductors.   Clearly, the transformer's secondary full-load current (72.12A) exceeds 30.8A.

(Again, referring to question No. 28. of Article 430 [Volume 3]) Based on a 60°C terminating means per Table 310.15(B)(16) as a minimum, the secondary conductors are required to be 2 AWG copper conductors which have a rated ampacity of 95A.

Considering condition **(2)** to NEC 240.21(C)(6), the secondary conductors must terminate in a single circuit breaker or set of fuses that limits the load current to not more than the ampacity of the secondary conductors. Based on the rated ampacity of the secondary conductors at 95 amps the conductors must be terminated in an overcurrent device rated no more than 90 amps. Being that the secondary conductors will terminate in an overcurrent device of the panelboard, the panelboard must be rated for 100 amps or higher per NEC 408.36.

To protect the panelboard

The *Exception* to NEC 408.36(B) states that when a panelboard is supplied by the secondary side of a transformer (such as in this question No. 26.) the panelboard shall be considered as protected by the overcurrent protection provided on the primary side of the transformer where that protection is in accordance with NEC 240.21(C)(1).

According to NEC 240.21(C)(1), single-phase (other than 2-wire) and multiphase (other than delta-delta, 3-wire) transformer secondary conductors are not considered to be protected by the primary overcurrent protective device (which is solely used to protect the transformer windings and not the transformers secondary conductors). As a result, for those transformers other than the above mentioned shall have overcurrent protection on the secondary side of a transformer and cannot be protected solely by the primary overcurrent device. This includes the 3-wire single-phase secondary conductors in this question (208/120V).

Conclusion

As it pertains to conditions **(1)** and **(2)** of NEC 240.21(C)(6) and assuming condition **(3)** is met, the following results are provided:

Transformer's primary overcurrent device: 40A
Transformer's secondary conductors: 2 AWG copper rated for 95A at 60°C
Transformer's secondary overcurrent device: No more than 90A

Part VIII (Supervised Industrial Installations)

240.91(B) - Devices Rated Over 800 Amperes (Protection of Conductors)

27. What size 3-phase feeder conductors are required to supply a 1200A main distribution panelboard with a like size main circuit breaker - if used in an automobile manufacturing facility where maintenance personnel are available 24-hours a day? Consider both copper and aluminum conductors rated for 75°C.

According to NEC 240.90, the provisions of NEC 240.91, which is inclusive of Part VIII of Article 240, shall be permitted only to apply to those portions of an electrical system in a supervised industrial installation that's used exclusively for manufacturing or process control activities.

NEC 240.91(B) states where an overcurrent device is rated over 800 amperes, the ampacity of the conductors it protects shall be equal to or greater than 95 percent of the rating of the overcurrent device specified in NEC 240.6 in accordance with related conditions. As a result, 1200A x .95 = 1140A.

As a minimum, the feeder conductors are required to have an ampacity not less than 1140 amperes without the consideration of being a continuous load. Per NEC 310.10(H), where the use of parallel conductors are specified and the application of two conductors per line is considered, each conductor must have an ampacity not less than 570A (1140A/2). Per Table 310.15(B)(16) at 75°C, two 1250 kcmil copper conductors rated for 590A each are required.

Per Table 310.15(B)(16), because a 2000 kcmil aluminum conductor carries a maximum ampacity of 560A at 75°C, three parallel conductor must be used instead where each conductor must have an ampacity not less than 380A (1140A/3). Therefore, three 750 kcmil aluminum conductors rated for 385A each are required.

240.92(B) - Feeder Taps (28. - 30.) [3]

According to NEC 240.92(B), feeder taps specified in NEC 240.21(B)(2), (B)(3), and (B)(4) are permitted to be sized in accordance with Table 240.92(B). Although derived from the original formula, the converted formulas applied in question Nos. 29. and 30. and the calculated results, are not considered proven methods for deriving such results. Each method was only used as an added approach to further enhance the reader's awareness of electrical calculations.

28. Considering the use of either the 350 kcmil or 500 kcmil copper secondary tap conductors that were determined in question No. 18., what size copper and aluminum secondary tap conductors would be required if the transformer's secondary could potentially produce 25,858 amps of short circuit current for a period of 5.16 seconds? Consider an initial temperature of the conductors before the short-circuit at 23°C (73.4°F) and an anticipated final temperature of the conductors upon termination of the short-circuit at 181°C (357.8°F).

Using the converted formulas below derived from the original formulas given in Table 240.92(B) to determine the conductor areas (C_A) in circular mils where,

$$C_A = \sqrt{\frac{t}{0.0297\log_{10}[(T_2 + 234)/(T_1 + 234)]}} \times I_{SC} \quad \textbf{(copper conductor)}$$

$$C_A = \sqrt{\frac{t}{0.0125\log_{10}[(T_2 + 228)/(T_1 + 228)]}} \times I_{SC} \quad \textbf{(aluminum conductor)}$$

$I_{SC} = 25,858A$

copper conductor

$$C_A = \sqrt{\frac{5.16s}{0.0297\log_{10}[(181 + 234)/(23 + 234)]}} \times 25,858A$$

$$= \sqrt{\frac{5.16s}{0.0297\log_{10}[(415)/(257)]}} \times 25,858A$$

$$= \sqrt{\frac{5.16s}{0.0297\log_{10}[1.615]}} \times 25,858A$$

$$= \sqrt{\frac{5.16s}{0.0297 \times .2082}} \times 25,858A$$

($\log_{10}[1.615] = .2082$ [rounded-off])

$$= \sqrt{\frac{5.16s}{.0062}} \quad \text{x } 25,858A$$

$$= \sqrt{832.26} \ (28.85) \ \text{x } 25,858A$$

$$= 746,003.3 \text{ cmil (or 746 kcmil [rounded-off])}$$

aluminum conductor

$$C_A = \sqrt{\frac{5.16s}{0.0125\log_{10}[(181+228)/(23+228)]}} \quad \text{x } 25,858A$$

$$= \sqrt{\frac{5.16s}{0.0125\log_{10}[(409)/(251)]}} \quad \text{x } 25,858A$$

$$= \sqrt{\frac{5.16s}{0.0125\log_{10}[1.629]}} \quad \text{x } 25,858A$$

$$= \sqrt{\frac{5.16s}{0.0125 \text{ x } .2119}} \quad \text{x } 25,858A$$

$$(\log_{10}[1.629] = .2119 \text{ [rounded-off])}$$

$$= \sqrt{\frac{5.16s}{.00265}} \quad \text{x } 25,858A$$

$$= \sqrt{1947.2} \ (44.13) \ \text{x } 25,858A$$

$$= 1,141,113.5 \text{ cmil (or 1141 kcmil [rounded-off])}$$

Refer to Table 8 of Chapter 9 to reference the circular mils area of selected tap conductors. Because the calculated results rendered a value other than a conductor of standard size the next standard size conductor was used. Where a copper tap conductor is used, 750 kcmil conductors (750,000 cmil) are required. Where an aluminum tap conductor is used, 1250 kcmil conductors (1,250,000 cmil) are required.

Although derived from the original formula, the converted formulas applied in question Nos. 29. and 30. and the calculated results, are not considered nor intended to be proven methods for deriving such results. The sole intent of each method only serves to further enhance the reader's awareness of performing other type electrical calculations.

29. Again per question No. 18., determine the amount of short-circuit current the 500 kcmil secondary tap conductors could experience if the short-circuit lasted 3.53s. Consider an initial conductor temperature of 28°C and a final conductor temperature of 237°C.

Using the converted formulas below derived from the original formulas given in Table 240.92(B) where,

$$I_{SC} = \sqrt{\frac{0.0297\log_{10}[(T_2+234)/(T_1+234)]}{t}} \quad \text{x A} \quad \textbf{(copper conductor)}$$

$$I_{SC} = \sqrt{\frac{0.0125\log_{10}[(T_2 + 228)/(T_1 + 228)]}{t}} \times A \quad \textbf{(aluminum conductor)}$$

500 kcmil conductor   A = 500,000 (conductor circular mil)

copper conductor

$$I_{SC} = \sqrt{\frac{0.0297\log_{10}[(237 + 234)/(28 + 234)]}{3.53s}} \times 500,000$$

$$= \sqrt{\frac{0.0297\log_{10}[(471)/(262)]}{3.53s}} \times 500,000$$

$$= \sqrt{\frac{0.0297\log_{10}[1.798]}{3.53s}} \times 500,000 \qquad (\log_{10}[1.798] = .2548 \text{ [rounded-off]})$$

$$= \sqrt{\frac{0.0297 \times .2548}{3.53s}} \times 500,000$$

$$= \sqrt{\frac{0.00757}{3.53s}} \times 500,000$$

$$= \sqrt{.00214} \; (.0463) \times 500,000$$

$$= 23,150A$$

aluminum conductor

$$I_{SC} = \sqrt{\frac{0.0125\log_{10}[(237 + 228)/(28 + 228)]}{3.53s}} \times 500,000$$

$$= \sqrt{\frac{0.0125\log_{10}[(465)/(256)]}{3.53s}} \times 500,000$$

$$= \sqrt{\frac{0.0125\log_{10}[1.816]}{3.53s}} \times 500,000 \qquad (\log_{10}[1.816] = .2591 \text{ [rounded-off]})$$

$$= \sqrt{\frac{0.0125 \times .2591}{3.53s}} \times 500,000$$

$$= \sqrt{\frac{0.00324}{3.53s}} \times 500,000$$

$$= \sqrt{.000918} \; (.0303) \times 500,000$$

$$= 15,150A$$

30. How much time will elapse if the 2 AWG conductors in question No. 16. experience a short-circuit current of 4589 amps where the initial conductor temperature of 33°C is increased to 133°C? Consider both copper and aluminum conductors.

Using the converted formulas below derived from the original formulas given in Table 240.92(B) where,

$$t = \frac{0.0297\log_{10}[(T_2 + 234)\,(T_1 + 234)]}{A^2/I^2 \text{ or } (A/I)^2} \text{ (copper conductor)}$$

$$t = \frac{0.0125\log_{10}[(T_2 + 228)\,(T_1 + 228)]}{A^2/I^2 \text{ or } (A/I)^2} \text{ (aluminum conductors)}$$

$I_{SC} = 4589A$ 2 AWG conductor   A = 66,360 (conductor circular mil)

copper conductor

$$t = \frac{0.0297\log_{10}[(133 + 234) \times (33 + 234)]}{(66,360 \text{ cmil} / 4589A)^2}$$

$$= \frac{0.0297\log_{10}[(367) \times (267)]}{(14.46)^2}$$

$$= \frac{0.0297\log_{10}[97,989]}{209.09}$$

$$= \frac{0.0297 \times 4.9912}{209.09}$$

$$= \frac{.1482}{209.09}$$

$$= .000709s \text{ or } .709 \text{ millisecond (ms)}$$

aluminum conductor

$$t = \frac{0.0125\log_{10}[(133 + 228)\,(33 + 228)]}{(66,360 \text{ cmil} / 4589A)^2}$$

$$= \frac{0.0125\log_{10}[(361) \times (261)]}{(14.46)^2}$$

$$= \frac{0.0125\log_{10}[94,221]}{209.09}$$

$$= \frac{0.0125 \times 4.9741}{209.09}$$

$$= \frac{.0622}{209.09}$$

$$= .000297s \text{ or } .297 \text{ millisecond (ms)}$$

ARTICLE 250 - Grounding and Bonding

Where the need for a grounded conductor existed in related questions of Volume 1 (Article 220) such conductor was used as a circuit conductor. In terms of application, this means that the grounded conductor would be utilized beyond the service equipment and therefore called, a *neutral* conductor. When the referenced grounded conductor of Article 250 is used as a neutral conductor, the size of the neutral conductor is then determined based on the provisions of Article 220.

However, when the grounded conductor is brought to the service equipment and used only as a means to ground a system and never extends beyond the service equipment the conductor is sized according to the provisions of Article 250.

Regardless, of whether used as a neutral conductor *or* as a grounded conductor, neither conductor is permitted to be sized smaller than an electrical system's grounding electrode conductor.

250.24(C)(1) - Sizing for a Single Raceway (Grounded Conductors Brought to Service Equipment) (1. - 6.) [6]

1.   A single-phase, 3W, 240/120 volts service is supplied with 1 AWG ungrounded copper conductors. Determine the minimum size grounded copper conductor needed per NEC.

According to NEC 250.24(C)(1), the grounded conductor must not be smaller than the required grounding electrode conductor as specified in Table 250.66. Based on the size of the 1 AWG ungrounded copper service conductors, a 6 AWG copper grounding electrode conductor is required. Therefore, the grounded conductor must be either equal to or larger than a 6 AWG copper conductor.

2.   For a 3-phase, 208/120V system, 600 kcmil aluminum conductors are being used as the service conductors. What is the minimum size grounded conductor allowed?

Referring to Table 250.66, either a 1/0 AWG copper or a 3/0 AWG aluminum conductor is the minimum size conductor allowed (compared to grounding electrode conductor) to serve as the grounded conductor based on the use of 600 kcmil aluminum service conductors.

Because copper and aluminum conductors differ in resistivity explains why a larger size aluminum conductor is required compared to that of a copper conductor to serve as the grounding electrode conductor.   As a result, both conductors practically share the same resistance in terms of cross-sectional area. If Table 8 of Chapter 9 is referenced you will see that an uncoated 1/0 AWG copper conductor (per 1000 feet) has a resistance of .122 ohm and a 3/0 AWG aluminum conductor almost share an equivalent resistance at .126 ohm.

3. A three-phase, 480/277V, 4W service is rated for 400 amps. Three 800 kcmil copper conductors will be used as the ungrounded service conductors. Determine the minimum size grounded conductor allowed for this service.

Again, referring to Table 250.66, either a 2/0 AWG copper or a 4/0 AWG aluminum conductor is the minimum size conductor allowed (compared to grounding electrode conductor) to serve as the grounded conductor when the service conductors are 800 kcmil copper conductors.

4.   Three 1250 kcmil copper conductors will supply a 600 amps service. As a minimum, what size grounded conductor is required? If three 2000 kcmil aluminum conductors were used, what size grounded conductor is required?

According to NEC 250.24(C)(1), when copper conductors are being used as service conductors and are larger than 1100 kcmil the grounded conductor is required to be not less than 12½ (12.5) percent (.125) of the area of the service conductors. Therefore, the grounded conductor is required to be 156.25 kcmil (1250 kcmil x .125), as a minimum. Since there is not a 156.25 kcmil conductor, the next standard size conductor must be used, which is, a 3/0 AWG (copper) conductor that has an area of 167,800 cmils or 167.8 kcmil according to Table 8 of Chapter 9. The use of a 3/0 AWG conductor will exceed the 12.5 requirement, (167.8/1250 = .1342 or 13.42 percent).

When aluminum conductors larger than 1750 kcmil are used as service conductor, again according to NEC 250.24(C)(1) the grounded conductor is required to be not less than 12.5 percent of the area of the service conductors. Therefore, the grounded conductor is required to be 250 kcmil (2000 kcmil x .125) as a minimum. A 250 kcmil aluminum or copper conductor is a standard size conductor.

In either situation, both minimum sized grounded conductors are at least the same size as the required grounding electrode per Table 250.66.

5. A service rated for 800 amperes will utilize 3-350 kcmil THW ungrounded copper conductors per phase to compensate for voltage drop. All ungrounded conductors will be installed in a common raceway. What size grounded conductor is required to supply this service?

Although, the service-entrance phase conductors are installed in parallel and in a common raceway, the requirements of NEC 250.24(C)(1) are still the same. The size of the grounded conductor (meaning one single conductor is needed) shall be based on the circular mil area of the largest set of parallel conductors, where in this situation are three identical sets and only one consideration needed. If this set of ungrounded service-entrance conductors exceeds the given conductors sizes, then the 12½ percent requirement must be implemented. Because of this, the three ungrounded parallel service conductors are required to be treated as one service conductor resulting to the following calculation,

$$350 \text{ kcmil x } 3 = 1050 \text{ kcmil}$$

As a minimum, based on Table 250.66 and the three parallel conductors having a single conductor equivalence of 1050 kcmil (where copper conductors over 600 through 1100 kcmil are referenced) either a 2/0 AWG copper or a 4/0 AWG aluminum (THW) grounded conductor is required to serve as the grounded conductor.

6. A 1600A, 3ϕ, 480/277V service will require the use of 4-600 kcmil ungrounded parallel copper conductors per phase. The conductor's insulation is rated for 75°C. An auxiliary gutter is being used to enclose the ungrounded conductors along with a service grounded conductor before terminating in service equipment. Determine the minimum size grounded copper conductor required for this service.

Because all conductors share a common enclosure and the four ungrounded parallel service conductors per phase are required to be treated as one service conductor per phase; in accordance with NEC 250.24(C) (1), where

$$600 \text{ kcmil x } 4 = 2400 \text{ kcmil}$$

which exceeds the maximum limit for copper service-entrance phase conductors (1100 kcmil) per Table 250.66, the 12.5 percent requirement of NEC 250.24(C)(1) must be applied to determine the size grounded conductor needed for this installation where,

$$2400 \text{ kcmil x } .125 = 300 \text{ kcmil}$$

As a minimum, at 75°C, a 300 kcmil copper grounded conductor is required for this service.

250.24(C)(2) - Parallel Conductors in Two or More Raceways (Grounded Conductors Brought to Service Equipment) (7. - .8) [2]

7. If the parallel ungrounded conductors in question No. 5. are now installed in three separate raceways, what size grounded service conductor is required to be installed in each separate raceway?

Because the ungrounded conductors are installed in separate raceways, each raceway must contain a separate grounded conductor that is based on the size of the ungrounded service-entrance conductor (meaning parallel conductors not treated as one conductor but individualized) in each raceway per NEC 250.24(C)(2).

Since 350 kcmil copper conductors will be installed in each separate raceway the separate grounded conductor that is required to be installed in each raceway is determined based on Table 250.66.

Table 250.66 identifies the use of a 2 AWG copper or 1/0 AWG aluminum conductor as the size conductor required to be used as the grounded conductor. However, NEC 250.24(C)(2) clearly states that the ungrounded service-entrance conductor in a raceway cannot be smaller than a 1/0 AWG conductor. Therefore, a 1/0 AWG copper conductor must be used instead of a 2 AWG copper conductor while the aluminum conductor can remain as is.

8. If the parallel ungrounded conductors in question No. 6. are now installed in four separate raceways, what size grounded service conductors are required to be installed in each separate raceway.

Because the service conductors will be installed in four separate raceways, NEC 250.24(C)(2) states, where installed in two or more raceways, the size of the grounded conductor in each raceway shall be based on the size of the ungrounded service-entrance conductor in the raceway but not smaller than 1/0 AWG. In accordance with Table 250.66, based on the use of 600 kcmil copper service (line) conductors, a 1/0 AWG copper conductor is referenced. As a result, for this installation a 1/0 AWG copper conductor must be installed in each raceway to serve as the grounded service conductor.

250.28(D) - Size (Main Bonding Jumper and System Bonding Jumper) (9. - 13.) [5]

What is a main bonding jumper? According to Article 100 a main bonding jumper is the connection between the grounded circuit conductor and the equipment grounding conductor at the service. NEC 250.24(B) states that for a grounded system, an unspliced main bonding jumper shall be used to connect the equipment grounding conductor(s) and the service-disconnect enclosure (metal panelboard, disconnect switch, switchboard, etc.) to the grounded conductor within the enclosure for each service disconnect. Abbreviated, NEC 250.28(A) - (D), requires the main bonding jumper to be made of copper or of a corrosion-resistant material where the jumper can be a wire, bus, screw, or similar suitable conductor (such as a strap). In the event the main bonding jumper is a screw, it must be green in color and visible when installed for inspection. And finally, the main bonding jumper must be attached per NEC

250.28(C). Equipment listed as suitable for service equipment is equipped with a main bonding jumper that must be field installed. Where sub-panelboards are involved the installation of a main bonding jumper is not allowed.

What purpose does the main bonding jumper serve? The main bonding jumper provides a low-impedance (little opposition) path for fault-current to clear a ground-fault by opening the overcurrent device that is relative to the faulted circuit. To sum things up the main bonding jumper is a very important component and plays an important role in a grounded system. See Figure 250.28(D).

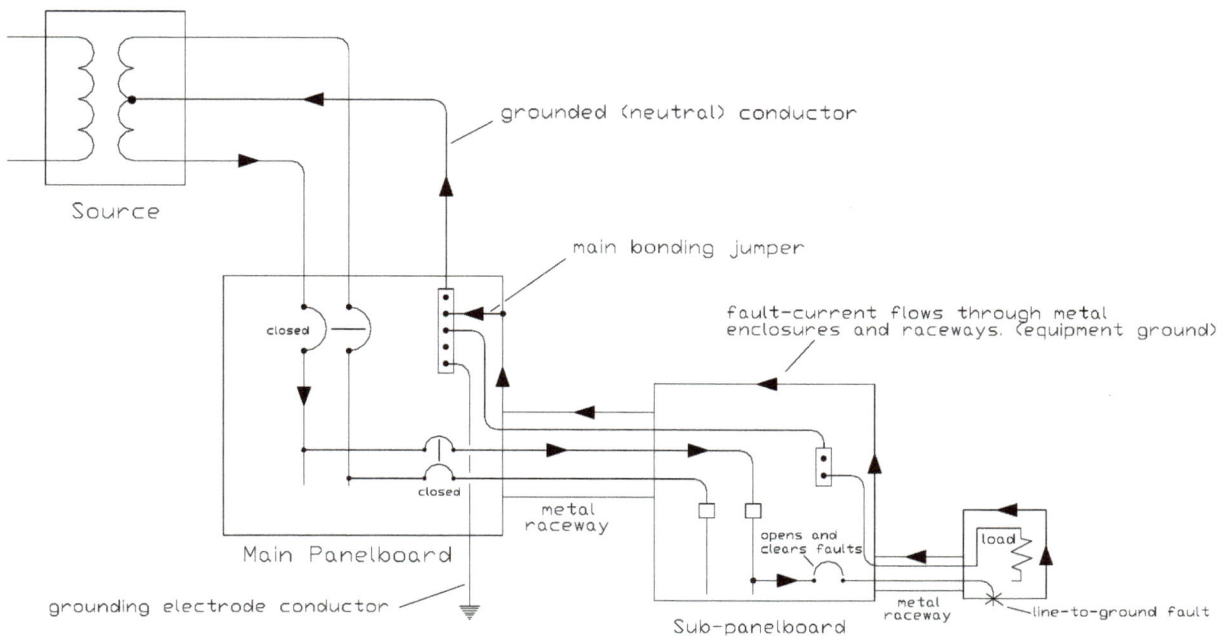

Figure 250.28(D)

With the main bonding jumper, a path is provided to the grounded conductor back to the source and eventually to the protective overcurrent device which opens the circuit, clears the fault and eliminates the possibilities of potential hazards. Without the main bonding jumper, the fault-current's path to the grounded conductor is disrupted thus allowing objectionable (unwanted, unintended, unacceptable) current and other hazardous conditions to exist (about the metal enclosures and raceways) without clearing the fault. As it pertains to the main bonding jumper and other grounding components, reference NEC 250.4(A)(5) and NEC 250.6(A). To sum it up, NEC 250.4(A)(5) requires an effective ground-fault current path for electrical equipment, wiring and other electrically conductive material whereas, NEC 250.6(A) requires associated grounding components to be installed and arranged in a manner that will prevent objectionable current.

9. A 200A single-phase 208/120 volts service will be supplied by 3/0 AWG THW copper service entrance conductors. What size copper main bonding jumper is required to bond the service enclosure and the equipment grounding conductors to the grounded service conductor?

According to NEC 250.28(D)(1), the main bonding jumper is sized according to Table 250.66. Based on 3/0 AWG copper service entrance conductors, a 4 AWG copper conductor (or larger) is required to use as the main bonding jumper.

10. Per Table 310.15(B)(7), 350 kcmil aluminum or copper-clad aluminum conductors can be used to supply a 300A service to a dwelling unit. If a copper conductor was used as a main bonding jumper, determine the minimum size conductor needed?

According to Table 250.66 a 350 kcmil aluminum conductor requires a 1/0 AWG aluminum or copper-clad aluminum main bonding jumper.

However, if a copper main bonding jumper is used, NEC 250.28(D)(1) states that when the phase conductors and main bonding jumpers are of different materials, the minimum size of the bonding jumper must be based on the assumed use of the phase conductors of the same material as the bonding jumper and with an ampacity equivalent to that of the installed phase conductors. In other words, the size of the phase conductors has to be reconsidered if using the same conductor material as that of the main bonding jumper.

In accordance with Table 310.15(B)(16) (at 75°C), since a 250 kcmil copper conductor (rated ampacity, 255A) is equivalent to a 350 kcmil aluminum conductor (rated ampacity, 250A) for a 300 amps service per Table 310.15(B)(7), the main bonding jumper has to be based upon a 250 kcmil copper conductor, where a 2 AWG copper main bonding jumper is required according to Table 250.66. As a result, a 2 AWG copper main bonding jumper is the minimum size bonding jumper that can be used with 350 kcmil aluminum service conductors.

11. Four 1750 THWN copper conductors will be used to supply a 700A, 3ϕ-4W, 440/254 volts service. Determine the minimum size copper main bonding jumper required to bond the required service components.

According to NEC 250.28(D)(1),

$$1750 \text{ kcmil x } .125 = 218.75 \text{ kcmil}$$

Referring to Table 8 of Chapter 9, a 250 kcmil copper conductor is the minimum size main bonding jumper that can be used per calculation.

12. A service rated for 1600 amps will be supplied by 4-600 kcmil parallel service conductors per phase. Based on the given parallel service conductors, determine the minimum size main bonding jumper required for the service equipment. The service conductors are copper and rated for 75°C.

Because there are four parallel conductors per phase the conductors must be treated as one conductor before proceeding therefore,

$$4 \text{ x } 600 \text{ kcmil} = 2400 \text{ kcmil (equivalent to one phase conductor)}$$

The main bonding jumper determined per NEC 250.28(D)(1) results to, 2400 kcmil x .125 = 300 kcmil. Use a 300 kcmil copper conductor rated for 75°C.

13. Referring to question No. 12., if an aluminum main bonding jumper is desired to be used with the copper service conductors instead, determine the minimum size of the bonding jumper.

The ampacity of a 600 kcmil copper conductor at 75°C is 420 amps. An aluminum conductor with an equivalent ampacity rated for 75°C is a 900 kcmil conductor (425 amps) which results to,

$$4 \text{ x } 900 \text{ kcmil} = 3600 \text{ kcmil x } .125 = 450 \text{ kcmil}$$

As a minimum, a 500 kcmil aluminum conductor (the next standard size) is required to be used as the main bonding jumper.

250.52(A) - Electrodes Permitted for Grounding (Grounding Electrodes)

14. A single-family dwelling requires the use a 300 amps service. The service conductors according to Table 310.15(B)(7) can be either 250 kcmil copper or 350 kcmil aluminum. According to Table 250.66 either a 2 AWG copper or a 1/0 AWG aluminum conductor can be used as the grounding electrode conductor. If the water pipes in this dwelling are nonmetallic, how is the grounding electrode conductor required to be installed?

If the water pipes in such dwelling are nonmetallic, NEC 250.50 requires one or more of the electrodes specified in NEC 250.52(A)(4) - (A)(7) to be installed. The electrode most often used in this case would be a ground rod or pipe per NEC 250.52(A)(5). As a result, the selected grounding electrode conductor is connected to the neutral (ground) bar of the service panel or the selected service panel where dual service panels are used and terminated on either a ground rod or pipe. Where a supplemental ground rod is used it must be bonded to either the ground rod or pipe (whichever is used) per NEC 250.53(A)(2) and 3, and NEC 250.53(B) and (E).

Table 250.64(D)(1) - Common Grounding Electrode Conductor and Taps (Service with Multiple Disconnecting Means Enclosures)

15. Five sets of THWN copper service entrance conductors are commonly connected to a service drop to supply five 480V, three-phase fusible disconnects to render a 1850A service. Each disconnect is rated for 75°C and are protected and served as follows:

DISCONNECT	FUSES	SERVICE CONDUCTORS
No. 1	200	3/0 AWG
No. 2	300	300 kcmil
No. 3	350	400 kcmil
No. 4	400	500 kcmil
No. 5	600	300 kcmil (2 paralleled)

What size grounding electrode conductor and individual tap conductors to the grounding electrode conductor are required for this installation? Use copper conductors.

NEC 250.64(D)(1) requires the grounding electrode conductor in such installation (commonly connected to a service drop/lateral) to be sized according to NEC 250.66 which reference Table 250.66. Because Table 250.66 is based on the largest ungrounded service-entrance conductors *or* the equivalent area for parallel conductors for sizing grounding electrode conductors, the circular mils area (using Table 8 of Chapter 9 where needed) for each individual and parallel conductors must be totaled per Table 250.66, **Note 1**. Table 250.66 must then be used to select a grounding electrode conductor per calculated circular mils area per service conductor.

SERVICE CONDUCTOR	CIRCULAR MILS (CM) AREA
3/0 AWG	167,800
300 kcmil	300,000
400 kcmil	400,000
500 kcmil	500,000
300 kcmil (2)	600,000
	1,967,800

Using the calculated total 1,967,800 cmils or 1967.8 kcmil, Table 250.66 (over 1100 kcmil) reference the use of a **3/0 AWG copper conductor** as the grounding electrode conductor for the combined service conductors. Unlike the provisions for grounded conductors where NEC 250.24(C)(1) requires the consideration of a 12½ (12.5) percent factor for service-entrance phase conductors larger than 1100 kcmil copper or 1750 kcmil aluminum the same does not apply to grounding electrode conductors.

Each individual tap conductor is also sized according to Table 250.66 based on the largest conductor serving each respective enclosure (disconnect).

SERVICE CONDUCTOR	TAP CONDUCTOR
3/0 AWG	4 AWG
300 kcmil	2 AWG
400 kcmil	1/0 AWG
500 kcmil	1/0 AWG
300 kcmil (2) @ 600 kcmil	1/0 AWG

NEC 250.64(D)(1) also requires each tap conductor to be connected to the grounding electrode conductor in such a manner that the grounding electrode conductor remains without a splice or joint per concluding methods (1) - (3).

Table 250.66 - Grounding Electrode Conductor for Alternating-Current Systems (16. - 22.) [7]

16. A single-family dwelling has a 125A service lateral. The service conductors supplying this service are 1/0 AWG aluminum USE. Determine the minimum size grounding electrode conductor (cold water ground) needed to ground the dwelling's metal water pipes.

NEC 250.66 states that the grounding electrode conductor must be sized according to Table 250.66 which requires an 8 AWG copper or a 6 AWG aluminum conductor to be used. Either type conductor can be used because the NEC makes no distinction as it pertains to the type (material) grounding electrode conductor that can be used.

17. A welding shop is required to have a 600A, 3ϕ-4W, 480/277V service to supply all calculated loads. The ungrounded service conductors being used are 1250 kcmil THW copper conductors. Determine the minimum size grounding electrode conductor needed to ground the shop's metal frame.

The grounding electrode conductor is required to be a 3/0 AWG copper or a 250 kcmil aluminum conductor. Where service conductors are sized over 1100 kcmil copper or over 1750 kcmil aluminum, the corresponding grounding electrode conductors are never required to be sized over 3/0 AWG copper or 250 kcmil aluminum.

18. Five 3/0 AWG THWN copper conductors are paralleled per phase to supply a 1000 amps service. What is the minimum size THWN copper conductor required to serve as the grounding electrode conductor?

The first requirement is to determine the circular mils area of a 3/0 AWG conductor. Referring to Table 8 of Chapter 9, the circular mils area of a 3/0 AWG is 167,800 cmils. Next determine the total circular mils area of all five parallel conductors,

$$167,800 \text{ cmils x } 5 = 839,000 \text{ cmils or } 839 \text{ kcmils}$$

Because the total circular mils of the five copper conductors totals 839 kcmil, in Table 250.66 it would fall between 600 and 1100 kcmil copper service conductors. Therefore, a 2/0 AWG copper grounding electrode conductor is required.

19. A 3ϕ - 4W, 150A service panelboard rated for 208/120V is located outside. The 1/0 AWG THWN copper conductors supplying the service can expect to be exposed to an ambient temperature of 112°F. Based on the given information, what size grounding electrode conductor is required where the major portion of the 150A service will supply nonlinear loads?

A 1/0 AWG THWN copper conductor has an ampacity of 150 amps. To adjust for ambient temperature the ampacity of the 1/0 AWG conductors have to be increased with respect to the applicable correction factors of Table 310.15(B)(2)(a). Because the grounded conductor of this voltage system is considered current-carrying per NEC 310.15(B)(5)(c), the correction factors of Table 310.15(B)(2)(a) and the application of Table 310.15(B)(3)(a) must be applied where four current-carrying conductors are considered, three ungrounded and one grounded. By applying the correction (.82) and adjustment (.80) factors, the required ampacity of the conductors is obtained,

$$\frac{150A}{.82 \text{ x } .80} = 228.66A \text{ (required ampacity)}$$

As a result, 4/0 AWG THWN copper conductors which are rated for 230A @ 75°C are required for the 150A service along with a 2 AWG grounding electrode conductor. Remember, Table 250.66 is based on the size of the largest ungrounded service-entrance conductor when sizing a grounding electrode conductor.

20. Three 2000 kcmil copper-clad aluminum conductors are being used to supply a 600 amps service. What size grounding electrode conductor is required?

A 250 kcmil aluminum conductor is the largest size grounding electrode conductor required regardless of size or parallel combinations of service entrance conductors. The same holds true for a 3/0 AWG copper grounding electrode conductor.

21. A 240/120V, 4W, 3φ service is supplied with two 400 kcmil copper line conductors, one 250 kcmil copper high leg (line) conductor and a 2/0 AWG copper neutral conductor. What size copper grounding electrode conductor is required for this installation?

Because the grounding electrode conductor is always selected based on the largest service-entrance conductor which is a 400 kcmil copper conductor, Table 250.66 reference the use of a 1/0 AWG copper grounding electrode conductor.

22. Three delta (Δ) connected transformers are used to supply a single-phase 240/120V, 3W service and a three-phase 240/120V, 4W service. The single-phase service is supplied by 4/0 AWG aluminum conductors and the three-phase service is supplied by 350 kcmil aluminum conductors. What size aluminum grounding electrode conductor is required for this installation?

Again Table 8 of Chapter 9 must be used to select a grounding electrode conductor per calculated circular mils area of each individual service conductor.

SERVICE CONDUCTOR	CIRCULAR MILS (CM) AREA
4/0	211,600
350 kcmil	350,000
	561,600

Using the calculated total 561,600 cmils or 561.6 kcmil, Table 250.66 (over 500 through 900) reference the use of a 3/0 AWG aluminum grounding electrode conductor for the two services.

250.102(C)(1) - Size for Supply Conductors in a Single Raceway or Cable (Size – Supply-Side Bonding Jumper) (23. - 24.) [2]

As a quick reference, the supply side of the service defines all service equipment associated with the service entrance conductors prior to the service entrance conductors terminating in the service disconnecting means (supply side - main breaker or fusible switch).

The equipment bonding jumper on the supply (line) side of the service equipment must not be smaller than the sizes listed in Table 250.66 for grounding electrode conductors. Equipment and main bonding jumpers are sized to handle the available fault-current that an electrical system is capable of supplying.

23. Three 500 kcmil THHW copper-clad aluminum conductors installed in metal raceway are being used as the service conductors for a 350A service. A grounding bushing is required for bonding the raceway to the service neutral/ground bus. What size copper-clad aluminum supply-side bonding jumper is required?

NEC 250.102(C)(1) reference Table 250.66 as the source for determining the size of an supply-side bonding jumper. A 1/0 AWG copper-clad aluminum conductor is the minimum size supply-side bonding jumper required for this installation.

24. An offset nipple is used to install 3-4/0 THWN copper service conductors from a meter can to a 225A panelboard with 75°C terminals. Because the nipple is inserted through the concentric knockout of the meter can the electrical continuity to ground is compromised and requires a bonding jumper based on NEC 250.92(B). Determine the minimum size copper supply-side bonding jumper required for this installation.

Per Table 250.66, a 2 AWG copper conductor rated for 75°C is the minimum size supply-side bonding jumper required.

250.102(C)(2) - Size for Parallel Conductor Installations (Size - Supply-Side Bonding Jumper) (25. - 27.) [3]

25. Three sets of 750 kcmil THWN-2 copper conductors installed in three individual raceways are being used to supply a 1600A 3-phase service. A single bonding jumper will be used to commonly bond each raceway. Determine the minimum size supply-side bonding jumper required to commonly bond the raceways.

Again referring to NEC 250.102(C)(1) as directed by NEC 250.102(C)(2), a common (single) bonding jumper must be based on the largest set of ungrounded supply conductors where,

$$750 \text{ kcmil} \times 3 = 2250 \text{ kcmil}$$
and
$$2250 \text{ kcmil} \times .125 = 281.25 \text{ kcmil}$$

Based on the calculated results, to commonly bond the raceways, a 300 kcmil copper conductor is needed to accommodate the 281.25 kcmil requirement.

If this same installation was being used to supply four individual 1200A service disconnecting means totaling 4800A; Table 250.66, **Note 1** (NEC 230.40, *Exception No.* 2) states that the equivalent size of the largest service-entrance conductor shall be determined by the largest sum of the areas of the corresponding conductors of each set. As an example, if Lines 1 and 2 of such 3-phase installation were being supplied by four 300 kcmil THWN-2 copper conductors while

Line 3 was being supplied by four 350 kcmil THWN-2 copper conductors the minimum size conductors required to commonly bond the raceways would be based on the total area of the 350 kcmil THWN-2 copper conductors. If calculated, per Table 8 of Chapter 9, a 4/0 AWG THWN-2 copper supply-side bonding jumper is required.

26. In question No. 25., if each individual raceway requires a separate supply-side bonding jumper, determine the minimum size copper conductor needed.

In this case, per NEC 250.102(C)(2), individual supply-side bonding jumpers are sized per Table 250.66 based on the size of the phase (ungrounded) conductors installed in each raceway.

Largest phase conductor = 750 kcmil

Separate 2/0 supply-side bonding jumper conductors are required.

27. This question relates to question No. 2. of Article 310. Refer to Figure 250.102(C)(2)-27. What size copper bonding jumpers (supply side) are needed to bond each individual metal raceway? What size copper bonding jumper is needed to commonly bond both parallel metal raceways?

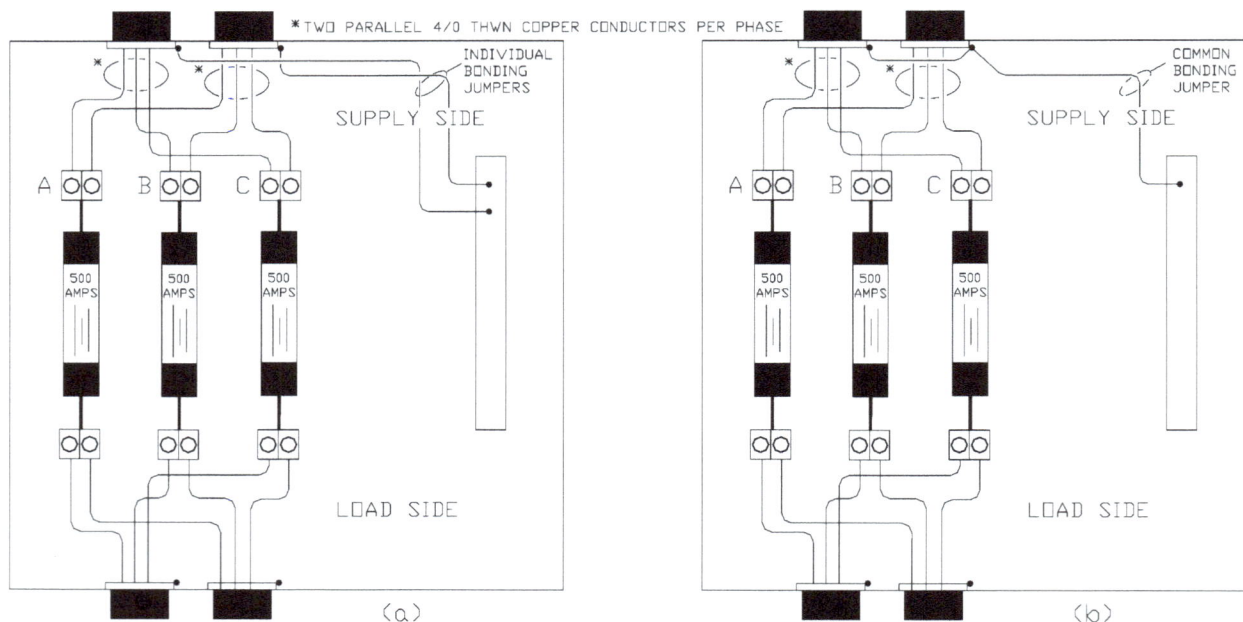

Figure 250.102(C)(2)-27

Individually Bond (a)
According to NEC 250.102(C)(2), where the ungrounded supply conductors are paralleled (two 4/0 conductors per phase) in two or more raceways (or cables), and an individual supply-side bonding jumper is used (one bonding jumper per raceway) for bonding the raceways (two individual raceways), the size of the supply-side bonding jumper for each raceway (or cable) shall be selected from Table 250.66 based on the size of the ungrounded supply conductors in each raceway (or cable).

Based on the 4/0 AWG copper service-entrance conductors installed in each raceway per question No. 2. of Article 310 and Figure 250.102(C)(2)-27, according to Table 250.66 *(over 3/0 through 350)*, a 2 AWG copper conductor is needed to bond each individual raceway.

Commonly Bond (b)
The last sentence of NEC 250.102(C)(2) requires a single (common) supply-side bonding jumper installed for bonding two or more raceways (or cables) to be sized in accordance with NEC 250.102(C)(1).

In recognizing the use of two parallel 4/0 THWN copper conductors per phase as service-entrance conductors, NEC 250.102(C)(1) states when such conductors are larger than 1100 kcmil copper or 1750 kcmil aluminum the bonding jumper on the supply side of a service shall have an area not less than 12½ percent of the area of the largest phase conductor.

Just as with all parallel installations, when two or more conductors are electrically joined at both ends, the conductors are identified as one single conductor. Being that the conductors in this case are sized in AWG dimensions opposed to kcmil (kilo circular mils) the size of the conductors has to first be converted to kcmil. Referring to Table 8 of Chapter 9 (Conductor Properties), a 4/0 AWG conductor has an equivalent circular mils (cm) area of 211,600 cm *or* 211.6 kcmil regardless of whether the conductor's material type is copper or aluminum. Considering there are two 4/0 AWG copper conductors serving as the service-entrance conductors, the equivalent (total) circular mils area between the two conductors is 423,200 cm (211,600cm x 2) or 423.2 kcmil. Clearly, in this case 423.2 kcmil does not exceed the 1100 kcmil requirement for a copper conductor per NEC 250.102(C)(1). As a result, the required size of the supply-side bonding jumper can be taken directly from Table 250.66. Based on the equivalent circulars mils area of the two parallel 4/0 AWG copper conductors *(over 350 through 600)*, a continuous 1/0 AWG copper conductor is required to commonly bond both parallel raceways.

250.102(C)(3) - Different Materials (Size – Supply-Side Bonding Jumper)

28. In question No. 24., if an aluminum bonding jumper is used based on the use of the 4/0 AWG copper service conductors, what size conductor is required?

Similar to NEC 250.28(D)(1), NEC 250.102(C)(3) has the same requirements when the phase conductors and supply-side bonding jumpers are of different materials. The approach for sizing an aluminum bonding jumper is the same as used in question No. 10.

Because a 300 kcmil aluminum conductor rated for 75°C is equivalent to a 4/0 AWG copper conductor ampacity-wise (230A), a 1/0 AWG aluminum conductor is the minimum size conductor that can be used as the supply-side bonding jumper per Table 250.66.

250.102(D) - Equipment Bonding Jumper on Load Side of an Overcurrent Device (29. - 32.) [4]

As a quick reference, the load side of the service defines all equipment after the service disconnecting means/overcurrent protection.

The equipment bonding jumper on the load side of an overcurrent devices shall be sized, as a minimum, in accordance with Table 250.122. Equipment bonding jumpers on the load side of the service disconnecting means/overcurrent protection are sized according to the overcurrent protective device ahead of all related circuit conductors and supporting equipment.

29. A metal conduit containing feeder conductors originates from a 600A main panelboard. The feeder conductors are protected by a 100 amps overcurrent device and the metal conduit requires being bonded to the panelboard. What size copper equipment bonding jumper is required?

NEC 250.102(D) reference Table 250.122 as the source for determining the size of an equipment bonding jumper on the load side of the service. Because the metal conduit encloses feeder conductors that are protected by a 100 amps overcurrent device, an 8 AWG copper conductor or a 6 AWG aluminum equipment bonding jumper is required.

30. The branch-circuit conductors of two individual circuits are protected by 15 and 20 amps circuit breakers. Each circuit is installed in ½" EMT. What size aluminum bonding jumpers are required to bond each raceway?

According to Table 250.122 a circuit protected by a 15A overcurrent device would require the EMT to be bonded with a 12 AWG aluminum conductor and the circuit protected by a 20A overcurrent device would require the EMT to be bonded with a 10 AWG aluminum conductor.

31. What size equipment bonding jumper would be required if both raceways in question No. 30. were required to be commonly bonded?

Per NEC 250.102(D), the equipment bonding jumper would be required to be based on the circuit having the largest overcurrent device and then sized according to Table 250.122. In this case, a 10 AWG aluminum conductor would be required or if a copper conductor is desired the use of a 12 AWG is permitted.

32. This question is related to question No. 2. of Article 310. Refer to Figure 250.102(D)-32. What size copper equipment bonding jumpers (load side) are needed to bond each individual metal raceway? What size copper equipment bonding jumper is needed to commonly bond both parallel metal raceways?

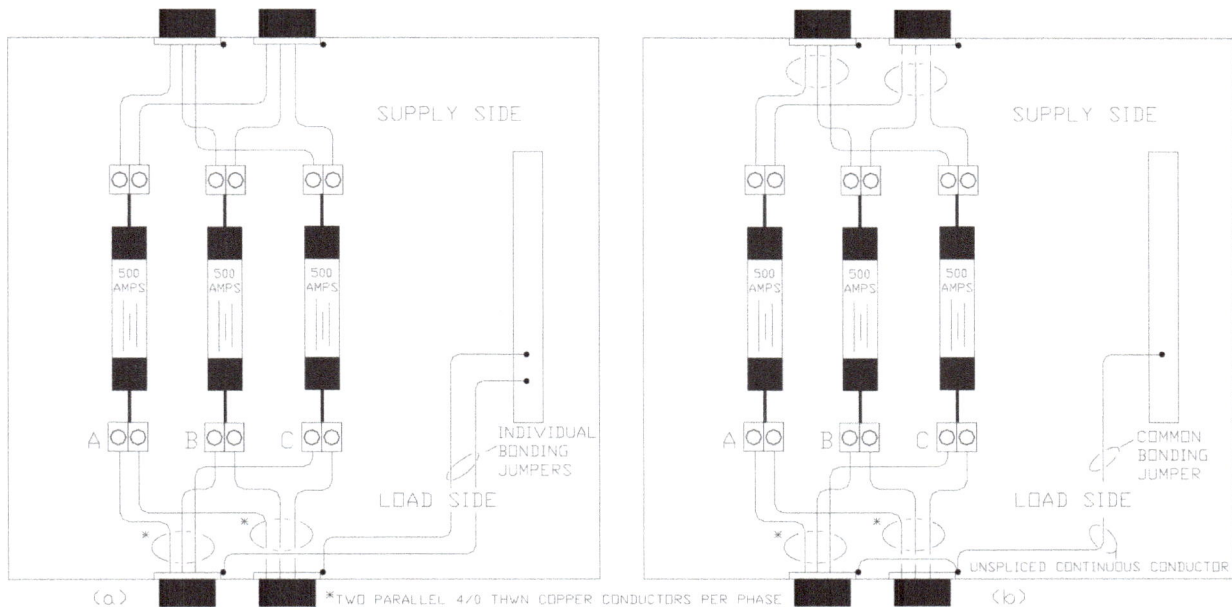

Figure 250.102(D)-32

Individually Bond (a)

Unlike the procedures required in NEC 250.102(C) for sizing an equipment bonding jumper on the supply-side of a service, NEC 250.102(D) in summary only requires an equipment bonding jumper on the load side of a service to be sized based on the provisions of Table 250.122. Where individual bonding jumpers are utilized to bond individual raceways, the size of the individual bonding jumper is based solely on the size of the overcurrent device ahead of (protecting) the ungrounded conductors which are enclosed by each individual metal raceway. Because the load side parallel 4/0 AWG conductors in Figure 250.102(D)-32(a) are protected by 500A fuses, each individual bonding jumper must be a 2 AWG copper conductor.

Commonly Bond (b)

Per NEC 250.102(D), a single common continuous equipment bonding jumper shall be permitted to connect two or more raceways where the bonding jumper is sized in accordance with Table 250.122 for the largest overcurrent device supply circuits therein. Based on the stated provision, the continuous bonding jumper shown in Figure 250.102(D)-31(b) which commonly bond the two metal raceways must also be a 2 AWG copper conductor per Table 250.122 where the 500A fuses are the largest overcurrent devices supplying the enclosed 4/0 AWG parallel conductors.

250.122(A) - General (Size of Equipment Grounding Conductors) (33. - 34.) [2]

33. What is the minimum size copper equipment grounding conductor required to be run along with copper feeder conductors from a 200A main panelboard to a 125A sub-panelboard, if the feeder is protected by a 100A overcurrent device?

If the feeder conductors are protected by a 100A overcurrent device, an 8 AWG copper equipment grounding conductor is required according to Table 250.122.

34. Referring to question No. 33., if an aluminum equipment grounding conductor is used instead, what size must the conductor be?

Again, according to Table 250.122, a 6 AWG aluminum conductor is required.

Also, refer to question No. 14. [392.60(B)] of Article 392 in reference to cable tray being used as an equipment grounding conductor.

250.122(B) - Increased in Size (Sizing of Equipment Grounding Conductors)

35. It is determined that three 3/0 AWG ungrounded copper feeder conductors supplying a 3φ service are inadequately sized to maintain a three percent voltage drop. The conductors are protected by a 200A overcurrent device located in a 1000A service panel. Because the service is located approximately 450' from its 208V source, 300 kcmil copper conductors must be used to compensate for the excessive voltage drop. What size equipment grounding conductor is required?

According to Table 250.122, the minimum size equipment grounding conductor required where the overcurrent protection is rated for 200 amps is a 6 AWG copper conductor. However, a 6 AWG conductor would prove to be inadequate because of the increase in size of the ungrounded conductors. Therefore, NEC 250.122(B) requires the equipment grounding conductor to be increased in size proportionately according to the circular mils (cmils) area of the ungrounded conductors. To meet these requirements, the following adjustments are required.

Refer to Table 8 of Chapter 9, to obtain the conductor's circular mils area.

$$3/0 \text{ AWG} = 167,800 \text{ cmils (initial size of ungrounded conductors)}$$
$$300 \text{ kcmil} = 300,000 \text{ cmils (increased size of ungrounded conductors)}$$
$$6 \text{ AWG} = 26,240 \text{ cmils}$$

$$\frac{300,000 \text{ cmils}}{167,800 \text{ cmils}} \times 26,240 \text{ cmils} = 46,913 \text{ cmils*}$$

*minimum size equipment grounding conductor required after adjustment

The calculated results reflect the proportional increase of the initial size equipment grounding conductor. The increased size equipment grounding conductor is based on the proportion of the increased to initial size ungrounded conductors which resulted to approximately 1.79 percent (300,000/167,800).

According to Table 8, a standard 3 AWG conductor (52,620 cmils) must be used based on the calculated increased (46,913 cmils) size of the equipment grounding conductor.

250.122(C) - Multiple Circuits (Sizing of Equipment Grounding Conductors)

36. Four individual circuits enclosed in the same nonmetallic raceway will share a common equipment grounding conductor. If the circuits are rated for 20A, 35A, 50A and 80A what is the minimum size copper equipment grounding conductor required?

The equipment grounding conductor must be sized based on the largest overcurrent device protecting circuit conductors in the same raceway. Because the circuit rated for 80 amps is the largest of the four, the size of the equipment grounding conductor is so determined. Although Table 250.122 does not list an 80A overcurrent device, the equipment grounding conductor must be based on the next overcurrent device listed in the table that exceeds 80A. Therefore, an 8 AWG (common) equipment grounding conductor is required based on a 100A overcurrent device.

250.122(D) - Motor Circuits (Sizing of Equipment Grounding Conductors)

37. A 3ϕ, 15hp motor rated for 460 volts has a 1.25 service factor and a nameplate current of 18 amps. The motor is being protected by a 175A instantaneous trip circuit breaker and an overload device rated for 22.5 amps. What size copper equipment grounding conductor is required?

In the 2005 NEC the equipment grounding conductor for a motor circuit was permitted to be sized based on the motor overload protective device. As it is now, NEC 250.122(D) listed two requirements for sizing the equipment grounding conductor for a motor circuit. To answer the question, NEC 250.122(D)(2) must be applied which requires such conductor to be sized per NEC 250.122(A) using the maximum permitted rating of a dual element time-delay fuse as the reference overcurrent device based on the provisions of NEC 430.52(C)(1), *Exception No. 1*.

NEC 430.52(C)(1) reference Table 430.52 to determine the size dual element time-delay fuse needed for this application. Per Table 430.52 such overcurrent device can be determined based on the use of the three-phase (polyphase) motor by increasing the motor's full-load current per Table 430.250 by 175 percent (1.75). In referring to Table 430.250 the full-load current for a three-phase, 460V motor is 21A. Using the following calculation where,

$$21A \times 1.75 = 36.75A$$

NEC 430.52(C)(1), *Exception No. 1* permits the use of a 40A dual element time-delay fuse based on the results of the calculated value.

Referring now to Table 250.122 the use of a 40A overcurrent device will require a 10 AWG copper equipment grounding conductor.

250.122(F) - Conductors in Parallel (Sizing of Equipment Grounding Conductors)

38. This question is affiliated with question No. 2. of Article 310. If used, what size equipment grounding conductors are required for the parallel raceway installation in Figure 250.122(F)-38?

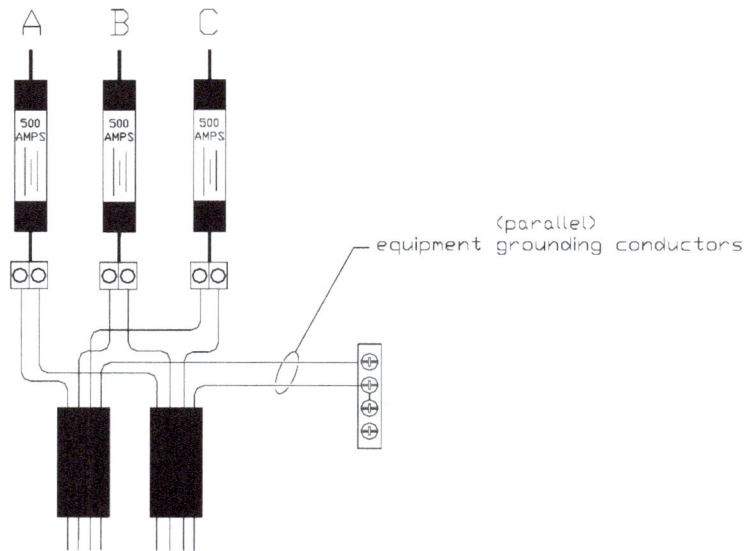

Figure 250.122(F)-38

According to NEC 250.122(F) when conductors are run in parallel in multiple raceway or cable, an equipment grounding conductor must be installed in each individual raceway. NEC 250.122(F) also states that each raceway must have an equipment grounding conductor sized according to the overcurrent device found in Table 250.122. NEC 310.10(H) also references the use of parallel equipment grounding conductors.

According to Table 250.122 the size of the equipment grounding conductors required to be installed in each individual raceway based on the size overcurrent device (500A fuses) protecting the parallel conductors per phase is either a 2 AWG copper or 1/0 AWG aluminum conductor.

250.122(G) - Feeder Taps (Sizing of Equipment Grounding Conductors)

39. If used, what size copper equipment grounding conductors are needed to accompany the 250 kcmil copper feeder conductors and the 6 AWG copper tap conductors in question No. 13. of Section 240.21(B)(1)?

Based on the provisions of NEC 250.122(G), the 300A overcurrent device protecting the 250 kcmil copper conductors per Table 250.122 reference the use of a 4 AWG copper conductor. However, since NEC 250.122(G) states that the equipment grounding conductor is not required to be larger than the tap conductors, a 6 AWG copper conductor can also be used.

ARTICLE 300 - Wiring Methods

300.34 - Conductor Bending Radius

1. A three-conductor cable containing individually shielded conductors are each .689 inches in diameter. The overall diameter of the cable is 2.21 inches. Determine the minimum bending radius required for the multiconductor cable.

Based on NEC 300.34, the minimum bending radius of an individually shielded conductor is 12 times the diameter of the conductor or 7 times the overall diameter of the cable, whichever is greater.

Minimum bending radius per conductor = .689" x 12 = 8.268"
Minimum bending radius per cable   = 2.21" x 7 = 15.47"

Based on the greater of the two, the minimum bending radius cannot be less than 15.47".

AMPACITY TABLES of ARTICLE 310

(Rated Up to and Including 2000 Volts)

TABLE	TEMP. RATING	CONDUCTORS	AMB. TEMP.
310.15(B)(16)	60°C-90°C (140°F-194°F)	COPPER/ALUM. (1)	30°C (86°F)
310.15(B)(17)	60°C-90°C (140°F-194°F)	COPPER/ALUM. (2)	30°C (86°F)
310.15(B)(18)	150°C-250°C (302°F-482°F)	COPPER/ALUM. (3)	40°C (104°F)
310.15(B)(19)	150°C-250°C (302°F-482°F)	COPPER/ALUM. (2)	40°C (104°F)
310.15(B)(20)	75°C-90°C (167°F-194°F)	COPPER/ALUM. (4)	40°C (104°F)
310.15(B)(21)	80°C (176°F)	BARE or COVERED	40°C (104°F)

(Rated 2001 to 35,000 Volts)

TABLE	TEMP. RATING	CONDUCTORS	AMB. TEMP.
310.60(C)(67)	90°C-105°C 194°F-221°F)	COPPER(5)	40°C (104°F)
310.60(C)(68)	90°C-105°C 194°F-221°F)	ALUMINUM(5)	40°C (104°F)
310.15(C)(69)	90°C-105°C 194°F-221°F)	COPPER(6)	40°C (104°F)
310.15(C)(70)	90°C-105°C 194°F-221°F)	ALUMINUM(6)	40°C (104°F)
310.60(C)(71)	90°C-105°C 194°F-221°F)	COPPER(7)	40°C (104°F)
310.60(C)(72)	90°C-105°C 194°F-221°F)	ALUMINUM(7)	40°C (104°F)
310.60(C)(73)	90°C-105°C 194°F-221°F)	COPPER(8)	40°C (104°F)
310.60(C)(74)	90°C-105°C 194°F-221°F)	ALUMINUM(8)	40°C (104°F)
310.60(C)(75)	90°C-105°C 194°F-221°F)	COPPER(9)	40°C (104°F)
310.60(C)(76)	90°C-105°C 194°F-221°F)	ALUMINUM(9)	40°C (104°F)
310.60(C)(77)	90°C-105°C 194°F-221°F)	COPPER(10)	20°C (68°F)
310.60(C)(78)	90°C-105°C 194°F-221°F)	ALUMINUM(10)	20°C (68°F)
310.60(C)(79)	90 C-105 C 194 F-221 F)	COPPER(11)	20°C (68°F)
310.60(C)(80)	90°C-105°C 194°F-221°F)	ALUMINUM(11)	20°C (68°F)
310.60(C)(81)	90°C-105°C 194°F-221°F)	COPPER(12)	20°C (68°F)
310.60(C)(82)	90°C-105°C 194°F-221°F)	ALUMINUM(12)	20°C (68°F)
310.60(C)(83)	90°C-105°C 194°F-221°F)	COPPER(13)	20°C (68°F)
310.60(C)(84)	90°C-105°C 194°F-221°F)	ALUMINUM(13)	20°C (68°F)
310.60(C)(85)	90°C-105°C 194°F-221°F)	COPPER(14)	20°C (68°F)
310.60(C)(86)	90°C-105°C 194°F-221°F)	ALUMINUM(14)	20°C (68°F)

(1) Not More Than Three Current-Carrying (In Raceway, Cable or Directly Buried)
(2) Single-Insulated (In Free Air)
(3) Not More Than Three Current-Carrying (In Raceway or Cable)
(4) Not More Than Three Single Insulated (Support on a Messenger)
(5) Insulated Single - Cables Triplexed (In Air)
(6) Insulated Single (In Air)
(7) Insulated - Three-Conductor Cable (Isolated in Air)
(8) Insulated - Triplexed or Three-Single Conductor Cables (Isolated Conduit - In Air)
(9) Insulated - Three -Conductor Cable (In Isolated Conduit - In Air)
(10) Three-Single Insulated (In Underground Electrical Ducts)
(11) Three Insulated Cabled - Three-Conductor Cable (In Underground Electrical Ducts)
(12) Single Insulated (Directly Buried)
(13) Three Insulated Cabled - Three-Conductor Cable (Direct Buried)
(14) Three Triplexed Single Insulated (Directly Buried)

Article 310, one of the most used Articles of the National Electrical Code underwent major changes in the 2011 edition. Although the scope and contents of this Article remains the same, numerical, format and added changes along with a few deletions has caused the entire Article to take on a totally different look. Therefore, beware of the challenges that these new changes may present.

ARTICLE 310 - Conductors for General Wiring

PARALLEL CONDUCTORS

NEC 310.10(H)(1) permits conductors or cables of sizes 1/0 AWG and larger to be connected in parallel for use as a single conductor. This type installation is based on 5 conditions, all of which must be met to satisfy the requirement. Once a combination of conductors is selected to satisfy such installation, the correct size raceway must then be considered.

When a raceway contains three or more current-carrying conductors the ampacity of the conductors must be adjusted. To avoid ampacity adjustments, it is best to install conductors in two or more separate raceways where applicable. NEC 300.20 requires all phase, grounded and equipment grounding conductors to be installed in the same raceway. Although, it is not necessary for neutral conductors and conductors of different phases to be the same as those of another phase, it is necessary that neutral conductors and those conductors of the same phase be of the same insulation type, length, material, cross-sectional area and terminate in the same manner. When equipment grounding conductors are required and there is more than one raceway involved, each individual raceway much includes a separate equipment grounding conductor. The equipment grounding conductor is sized according to NEC 250.122 and is determined by the ampere rating of the overcurrent protective device that protects the conductors in the raceway or cable.

SIZING PARALLEL CONDUCTORS

NEC 310.10(H)(1) provides conditions when feasible to replace a very large conductor with two or more smaller conductors that are connected in parallel to form one single conductor. In the study of Electrical Theory, when a set of two or more resistors are connected in parallel, the total resistance of the set is less than the resistance of the smallest resistor. By connecting electrical conductors in parallel more paths are created thus allowing more current to flow among the conductors because of the decrease in resistance produced by the parallel connection. However, the flow of current must be evenly divided between the parallel conductors, that is, if two parallel conductors are to receive a total of 50 amps, each conductors must receive half of the total (25 amps) and one-third of the total (16.67 amps) for three parallel conductors and so forth and so on.

To evenly direct the flow of current between parallel conductors, the conductors must be made of the same material, measure the same in length, have the same cross-sectional area, be of the same insulation type and terminate in the same manner. Along with this, conductors that are run in parallel must be at least 1/0 AWG or larger in size, excluding *Exceptions*.

Smaller conductors have a much higher ampacity per cross-sectional area (kilo circular mils - kcmil) than larger conductors. Thus, when smaller conductors are run in parallel instead of using one larger conductor, the combined cross-sectional area of the smaller conductors, depending on the combination, is generally less than the cross-sectional area of the larger conductor. By installing conductors in parallel the cost of installation can be significantly reduced, nevertheless adjustments may be required to compensate for higher voltage drops due to a decrease of total resistance created by such installation.

NEC 310.15(B)(3) must be adhered to when parallel conductors, installed in the same raceway, consist of four or more current-carrying conductors. When separate raceways are used to install parallel conductors, NEC 300.20(A) requires all phase conductors along with grounded and equipment grounding conductors if used, to be grouped together. Raceway enclosing parallel conductors must be of the same physical characteristics, which means, raceways of different types cannot be used together.

When phase conductors, polarity-wise (DC [+/-]) conductors, neutral conductors, grounded circuit conductors or equipment grounding conductors are installed in parallel the following guidelines are provided based on the provisions of NEC 310.10(H)(2) [See Figure 310.10(H)(2)]:

(1) Each set of parallel conductors of the same phase must be of the same length

(2) Each set of parallel conductors of the same phase must be of the same material (that is, copper, aluminum or copper-clad aluminum)

(3) Each set of parallel conductors of the same phase must be the same size (in circular mil area)

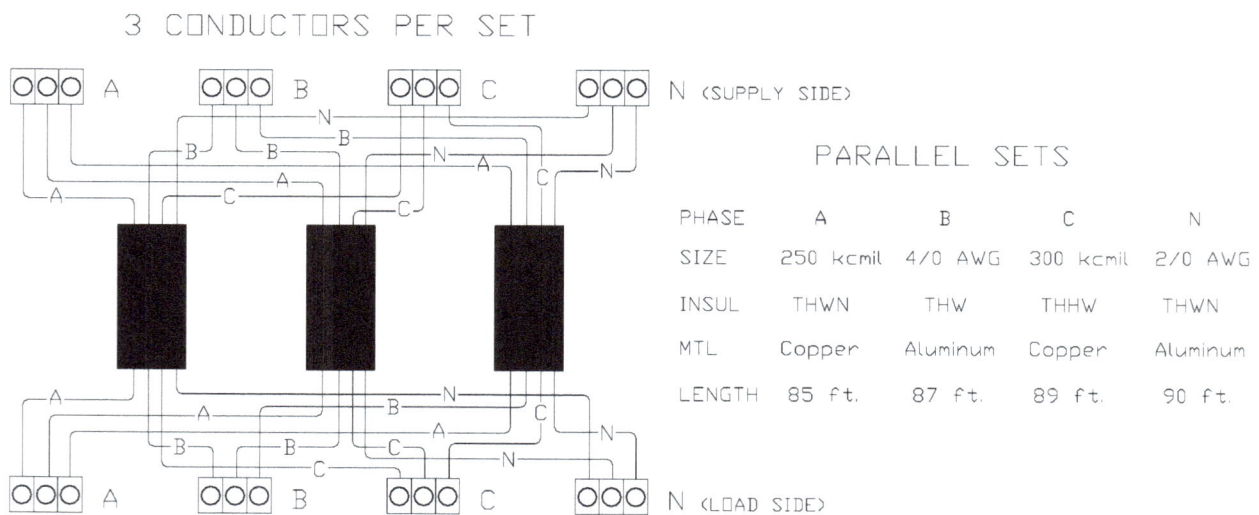

PHASE	A	B	C	N
SIZE	250 kcmil	4/0 AWG	300 kcmil	2/0 AWG
INSUL	THWN	THW	THHW	THWN
MTL	Copper	Aluminum	Copper	Aluminum
LENGTH	85 ft.	87 ft.	89 ft.	90 ft.

Figure 310.10(H)(2) - Installation requirements for parallel conductors

(4) Each set of parallel conductors of the same phase must be of the same insulation type

(5) Each set of parallel conductors of the same phase must be terminated in the same manner

(3) When installed in separate raceways or cable the raceway or cable must contain the same number of conductors and be of the same electrical characteristics (that is, all EMT, rigid metal, PVC, or other similar type raceways or either all steel-sheathed, aluminum-sheathed or similar type cables. The mixing of raceways or cables (types) is not allowed).

Conductors of one phase, polarity, neutral, grounded circuit conductors or equipment grounding conductors are not required to have the same physical characteristics as those of another phase, polarity, neutral, grounded circuit conductors or equipment grounding conductors to achieve balance. The intent here is not for all parallel conductors of different phases to be of the same conductor material that is, all copper, aluminum or copper-clad aluminum. However, the intent is for all parallel conductors of different phases to be of the same length, conductor material, size, insulation type and terminate in the same manner as illustrated in Figure 310.10(H)(2).

(4) Conductors installed in parallel shall comply with the provisions of NEC 310.15(B)(3)(a), *More Than Three Current-Carrying Conductors in a Raceway or Cable.*

(5) Equipment grounding conductors where needed, that are installed with parallel conductors must be sized in accordance with NEC 250.122.

(6) Parallel equipment bonding jumpers installed in raceways shall be sized and installed in accordance with NEC 250.102.

insulation - The outer material which encloses a bare conductor that serves to keep current in its intended path that is recognized by the National Electric Code.

310.10(H)(1) - General (Conductors in Parallel) (1. - 2.) [2]

1. What size THWN copper conductors are required to supply a 3-wire 3-phase system that's protected by 500A fuses?

Referring to Table 310.15(B)(16) there are no THWN copper conductors that has an ampacity of exactly 500 amperes. An 800 kcmil THWN copper conductor has an ampacity of 490 amperes and a 900 kcmil THWN copper conductor has an ampacity of 520 amperes. For certain, the 900 kcmil conductor can be used because the ampacity of the conductor will never be exceeded if protected by 500A fuses. As for the 800 kcmil conductor, its ampacity will be exceeded if protected by the same 500A fuses, but does that mean it can't be used.

According to NEC 240.4(B) if the ampacity (490A) of the 800 kcmil THWN copper conductor does not match the standard rating of an overcurrent device, where in this case the 500A fuses, and the rating of the fuses does not exceed 800 amperes, the next standard size fuses above the ampacity of the conductors can be used providing the first condition of NEC 240.4(B) is also met. Therefore, 800 kcmil THWN copper conductors can be used and protected by 500A fuses.

Refer to NEC 240.6(A) to review standard ampere ratings of fuses and fixed-trip circuit breakers.

2. If two parallel conductors (THWN-copper) per phase were required to serve the same 3-phase system in question No. 1., determine the size of the conductors if protected by the 500A fuses.

When the need to determine the size of parallel conductors occurs, the following formula can be used to calculate the minimum required ampacity of the parallel conductors to determine the size of needed conductors,

$$\text{Required Ampacity} = \frac{\text{Conductor's Overcurrent Protection}}{\text{Parallel conductors per phase x Derating factors (if needed)}}$$

Since the parallel conductors will consist of two conductors per phase, let's start by dividing the 500A fuses by 2 to determine the required ampacity of the parallel conductors.

$$\frac{500A}{2} = 250A$$

Observe Figure 310.10(H)(1)-2. Figure 310.10(H)(1)-2 is an illustration of the installation described in question No. 2. Take notice of the individual raceway (assumed to be metal) enclosing each set of parallel conductors. Because each raceway will enclose only 3 current-carrying conductors and no other adverse conditions are stated, ampacity derating is not required.

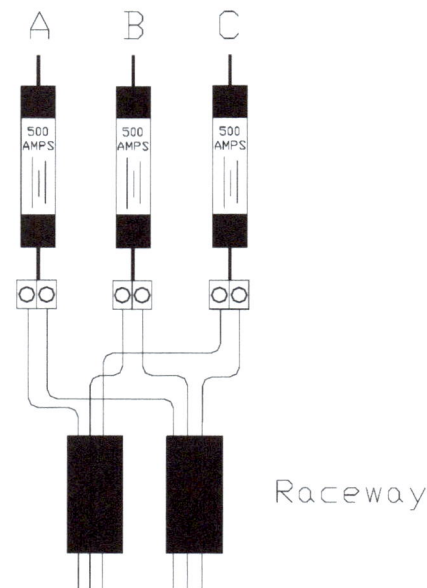

Figure 310.10(H)(1)-2

Again, referring to Table 310.15(B)(16), we find there are no THWN copper conductors that have an ampacity of exactly 250 amperes. A 4/0 AWG THWN copper conductor has an ampacity of 230 amperes and a 250 kcmil THWN copper conductor has an ampacity of 255 amperes.

According to NEC 310.10(H)(1), parallel conductors are permitted to be connected in parallel for the purpose of serving as one single conductor. Therefore, if the two 4/0 AWG conductors are used they will combine to produce a total ampacity of 460 amperes compared to two 250 kcmil conductors which will produce a total ampacity of 510 amperes.

With the exception of using two parallel conductors per phase opposed to single conductors, question No. 2. is quite similar to question No. 1. Since there are no standard size fuses listed in NEC 240.6(A) with a rating between 450A and 500A, the next standard size set of fuses to protect the 4/0 AWG conductors are the 500A fuses. As a result, two parallel 4/0 AWG THWN copper conductors can be used per phase instead of two parallel 250 kcmil THWN copper conductors based on the provisions of NEC 240.4(B).

310.15(A)(2) - Selection of Ampacity (General)

3. Before terminating in a panelboard which primarily feeds welding circuits, a 240V, 3W feeder is enclosed in a 2" raceway which is routed through three separate areas where the ambient temperatures are 24°C, 40°C and 106°F. If the feeder consists of three 250 kcmil THWN copper conductors, determine the ampacity of the conductors.

NEC 310.15(A)(2) requires the lowest (calculated) value to be used when more than one ampacity applies for a given circuit length.

Table 310.15(B)(16) lists the rated ampacity of a 250 kcmil THWN copper conductor at 255 amps.

The correction factors of Table 310.15(B)(2)(a) list the following multipliers per ambient temperature for a THWN copper conductor,

$$1.05 @ 24°C \quad .88 @ 40°C \quad .82 @ 106°F$$

Without having to perform three separate calculations to determine the allowable ampacity of the 250 kcmil copper conductors, it should be obvious that the lowest calculated value would result from the correction factor with the lowest multiplier, .82. Therefore,

$$255A \times .82 = 209.1A$$

The ampacity of the 250 kcmil THWN copper conductors is limited to 209.1A, the lowest calculated value.

310.15(A)(2), *Exception* - Selection of Ampacity (General)

4. Refer to the Figure 310.15(A)(2)-4. A 600A, 3ϕ motor control center rated for 600VAC is being fed from a 480V, 3ϕ switchgear compartment. The distance between the switchgear and the motor control center and means of supply are as shown. Determine the allowable ampacity of the feeder conductors.

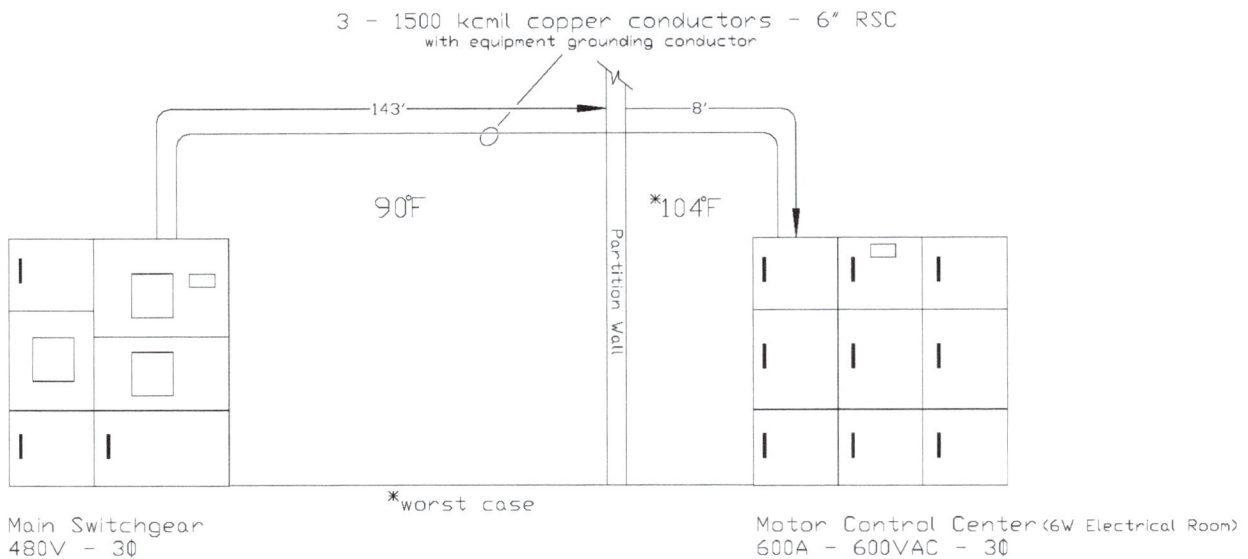

Figure 310.15(A)(2)-4

The *Exception* to NEC 310.15(A)(2) provides an alternate means to the rule other than having to use the lowest ampacity value when more than one ampacity applies for a given circuit length.

Paraphrased, the *Exception* states where two different ampacities apply to adjacent (adjoining) portions of a circuit, the higher ampacity (of the two ampacities of adjacent portions) shall be permitted to be used beyond the point of transition (surpassed), providing the surpassed distance is 10 feet or less or the distance is no more than 10 percent of the conductor's high ampacity length that surpasses the point of transition. In either case, the lesser of the two must be used.

Per Figure 310.15(A)(2)-4, three 1500 kcmil copper conductors enclosed in 6" rigid steel conduit are routed through adjacent areas where the ambient temperatures are 90°F and 104°F. Considering the ambient temperatures, the allowable ampacity of the feeder conductors must be determined based on the applicable correction factor of Table 310.15(B)(2)(a). Therefore,

at   90°F - conductor allowable ampacity = 587.5A (625A x .94)
at 104°F - conductor allowable ampacity = 550A (625A x .88)

Now let's consider the two conditions,

(1) at a distance that is 10 feet or less
The final length of the circuit is transitioned from an area where the ambient temperature is 90°F and terminates in an area where the ambient temperature is 104°F. Because the terminating distance of the conductors is less than 10 feet, the ampacity of the conductors is permitted to be the same as that calculated at 90°F (587.5A), although the conductors are in an area where the ambient temperature is 104°F.

(2) at a distance that is 10 percent (10 percent or .10) of the circuit's entire length
Ten percent of the conductor's high ampacity length (143') that surpasses the point of transition is 14.3 feet (143 feet x .10). As a result, the ampacity calculated at 90°F (587.5A - the high ampacity) can also be applied up to 14.3 feet. However, any length over 14.3 feet will require the conductor's ampacity to be limited to the ampacity calculated at 104°F (550A).

Without the *Exception*, the rule would have required the lowest value to be used regardless.

310.15(A)(3) - Temperature Limitation of Conductors

5. A 1" conduit contains the following current-carrying conductors:

2-8 AWG THW copper   4-10 AWG TW copper   3-12 AWG THWN-2 aluminum

If the conduit is routed through an area where the ambient temperature at times reaches 106°F, determine the allowable ampacity of each conductor.

This question serves as an example of how the provisions of NEC 310.15(A)(3) are applied. Conductors are never to be used under any conditions that will cause its insulation to reach a temperature that is greater than the temperature for which it was designed for.

The conductor temperature rating of the 8 AWG THW copper conductor = 75°C
The conductor temperature rating of the 10 AWG TW copper conductor = 60°C
The conductor temperature rating of the 12 AWG THWN-2 aluminum conductor = 90°C

Since the 10 AWG TW copper conductor has the lowest temperature rating the ampacity of each conductor must be based on 60°C.

Based on an ambient temperature of 106°F, the correction factor for both copper and aluminum conductors at 60°C is .71 per Table 310.15(B)(2)(a).

The total number of current-carrying conductors in the raceway (conduit) is 9, based on 9 current-carrying conductors per Table 310.15(B)(3)(a), the adjusted factor is .70.

The ampacity of the 8 AWG THW copper conductor = 40 amps at 60°C
The ampacity of the 10 AWG TW copper conductor = 30 amps at 60°C
The ampacity of the 12 AWG THWN-2 aluminum conductor = 15 amps at 60°C

Both factors are then multiplied by the ampacity of the each conductor which results in the allowable ampacity of each conductor.

(8 AWG THW) 40 amps x .71 x .70 = 19.88 amps
(10 AWG TW) 30 amps x .71 x .70 = 14.91 amps
(12 AWG THWN-2) 15 amps x .71 x .70 = 7.46 amps

Conductors having a higher conductor (insulation) temperature rating are designed to operate at higher temperatures. Because the ampacity of these conductors are based on their temperature ratings they are capable of safely carrying more current than those conductors having a lower temperature rating. Since the amount of current flowing through a conductor is proportional to the amount of heat the conductor will produce, a conductor operating at its maximum rated ampacity will produce a proportional amount of heat. This heat will add to the surrounding (ambient) temperature. For the conductor having the higher temperature rating this is safe, yet unsafe for the conductors with the lower temperature rating. Because of this, the ampacity of the conductors having higher temperature ratings are required to be reduced to eliminate unsafe operating conditions for conductors having lower temperature ratings.

Table 310.15(B)(16) - Allowable Ampacities of Insulated Conductors Rated Up to and Including 2000 Volts, 60°C Through 90°C (140°F - 194°F), Not More Than Three Current-Carrying Conductors in Raceway, Cable, or Earth (Directly Buried), Based on Ambient Temperature of 30°C (86°F)

WHAT DETERMINES A CONDUCTOR'S AMPACITY?

For clarity, let's first recognize that the term *wire*, which is used for most in the electrical profession, is hardly ever used in the National Electrical Code (NEC). As referenced, almost every section of the NEC where applied use the term *conductor* instead which could actually be a wire (as referenced in NEC 376.2, Tables 4, 5, 5A and Informative Annex C) *or* some other type metallic component deemed suitable for carrying current safely such as a busbar. In reality all wires are conductors yet all conductors are not wires. Case in point, a busbar is a conductor but not a wire. With this in mind we can now proceed.

When current flows through a conductor it produces heat. There are limits as to the amount of heat various types of insulation can safely withstand; even bare conductors must not be allowed to reach unsafe temperatures. Table 310.104(A) [rated for 600 Volts] and Tables 310.15(B)(16) through 310.15(B)(20) [rated up to and including 2000 Volts] of the NEC list the temperature ratings of the insulation that encloses a conductor. The temperature rating of a conductor's insulation is realized when a conductor carries its total ampacity and operates within a specified ambient temperature. If the insulation of a conductor has a temperature rating of 167°F (75°C) this does not mean that the conductor may be used where the ambient temperature is 167°F (75°C). Rather, it means that the rated temperature of the conductor's insulation must not exceed 167°F (75°C). The conductor's insulation is exposed to 167°F (75°C) when the conductor operates at its rated ampacity in an area where the ambient temperature is 86°F (30°C). If the area which the conductor is installed exceeds the 86°F (30°C) ambient temperature and the conductor operates at its rated ampacity; this will expose the conductor's insulation to a temperature above its 167°F (75°C) rating. In this case the ampacity of the conductor must be reduced.

Excessive heat can damage the insulation of a conductor in various ways. This depends on the type of material the insulation is made of and the amount of heat the insulation is exposed to. When overheated, some insulation will soften, melt, or harden resulting in breakdown. Once breakdown occurs it can cause ground faults, short circuits, fires and possible personal injuries or fatalities. Therefore, it is safe to say, "heat is insulation's worst enemy".

The source of this damaging heat originates from the ambient temperature and the current which flows through the conductor. Heat varies as the square of current, thus heat created in a conductor can be calculated by the formula,

$$\text{Watts} = I^2R$$

The more current-carrying conductors installed in a raceway or where single conductors or multiconductor cable are grouped and exceeds 24 inches in length, the more damage the heat produced by the conductors can cause. Conductors that are directly buried, installed in raceway, or in the form of a multiconductor cable, cannot dissipate heat caused by current flowing through them as freely as a conductor installed in free air. Because of this, heat must be limited and held to an acceptable level that is safe. This process is achieved by restricting the number of conductors installed in a raceway or providing adequate spacing for grouped single conductors or multiconductor cable and last, limiting the amount of current a given conductor is capable of carrying. As a result, certain underlined derating factors must be applied towards reducing the ampacity of a conductor.

derating factor - A factor used to reduce the ampacity of a conductor when a conductor is used in an adverse environment, or when more than the allowed number of current-carrying conductors (installed in cable or in raceway or directly buried) are exceeded.

UNDERSTANDING TABLE 310.15(B)(16) formerly TABLE 310.16

Tables 310.15(B)(16) through **310.15(B)(20)** are all basically the same in structure and format. Since **Table 310.15(B)(16)** is mostly used and referenced for most electrical installations it will be used as the sole reference for all explanations and examples.

For those who are not familiar or experience difficulty using **Table 310.15(B)(16),** each applicable column of the table has been identified followed by a brief explanation to enhance the understanding of the table.

Table 310.15(B)(16) (formerly Table 310.16) Allowable Ampacities of Insulated Conductors Rated Up to and Including 2000 Volts, 60°C Through 90°C (140°F Through 194°F), Not More Than Three Current-Carrying Conductors in Raceway, Cable, or Earth (Directly Buried), Based on Ambient Temperature of 30°C (86°F)*

(C) Size AWG or kcmil	(A) Temperature Rating of Conductor [See Table 310.104(A).]						(C) Size AWG or kcmil
	60°C (140°F)	75°C (167°F)	90°C (194°F)	60°C (140°F)	75°C (167°F)	90°C (194°F)	
	(B) Types TW, UF	Types RHW, THHW, THW, THWN, XHHW, USE, ZW	Types TBS, SA, SIS FEP, FEPB, MI, RHH, RHW-2, THHN, THHW, THW-2, THWN-2, USE-2, XHH, XHHW, XHHW-2, ZW-2	Types TW, UF	Types RHW, THHW, THW, THWN, XHHW, USE	Types TBS, SA, SIS, THHN, THWN, THW-2, THWN-2, RHH, RHW-2, USE-2, XHH, XHHW, XHHW-2, ZW-2	
	(D) COPPER			(D) ALUMINUM OR COPPER-CLAD ALUMINUM			
12**	(E) 20	25	30	15	20	25	12**
6	55	65	75	40	50	55	6
1	110	130	145	85	100	115	1
4/0	195	230	260	150	180	205	4/0
500	320	380	430	260	310	350	500
750	400	475	535	320	385	435	750
1500	525	625	705	435	520	(E) 585	1500
2000	555	665	750	470	560	630	2000

(F) *Refer to 310.15(B)(2) for the ampacity correction factors where the ambient temperature is other than 30°C (86°F).
**Refer to 240.4(D) for conductor overcurrent protection limitations.

Table 310.15(B)(16)

(A) - Temperature ratings of the insulation enclosing the conductors [60°C(140°F), 75°C(167°F), and 90°C(194°F)]

(B) - Type insulation enclosing the conductors [See Table 310.104(A) - Conductor Application and Insulations Rated 600V]

(C) - Size of the conductors [Given in American Wire Gages (AWG) (sizes 18 - 4/0) and kilo-circular mils (kcmil) (sizes 250-2000)]

(D) - Type material the conductors are made of (Copper, Aluminum, Copper-clad Aluminum)

(E) - The ampacity of a given conductor based on size, type of material the conductor is made of, and temperature rating of conductor

(F) - Footnote asterisk (*) reference the use of new NEC section 310.15(B)(2) for ampacity correction factors when ambient temperature is other than 30°C (86°F)
Footnote asterisk (**) reference NEC 240.4(D) which identifies the protection of conductors and the allowable overcurrent protection for selected conductors.

ambient temperature - Temperature of air that surrounds an object on all sides.

ampacity - The amount of current a conductor can safely carry continuously under the conditions of use without exceeding its temperature rating.

Table 310.104(A) - Conductor Application and Insulations - This table contains information on the various types of insulation conductors are enclosed and protected by. It also provides applicable data pertaining to the use and application of a conductor.

temperature rating - The maximum permitted temperature a conductor's insulation is designed or rated to operate continuously under specific conditions.

Before concluding, in summary, Table 310.15(B)(16) is referenced when the ampacity of a conductor is needed per temperature rating (60°C, 140°F....), insulation type (TW, THWN...) and type material (copper, aluminum....). When specific information is needed pertaining to the type insulation that encloses a conductor, Table 310.104(A)(Conductor Applications and Insulations Rated 600 Volts) is then referenced. For detailed information pertaining only to the core (bare) conductor (without insulation), Table 8 (Conductor Properties) of Chapter 9 is referenced.

UNDERSTANDING TABLE 310.15(B)(2)(a)
formerly CORRECTION FACTORS to TABLE 310.16

Table 310.15(B)(2)(a) is one of the major changes that occurred in Article 310 of the 2011 NEC. This table was once a part of former Table 310.16 which is now Table 310.15(B)(16). The correction factors of **Table 310.15(B)(2)(a)** are applied when ambient temperatures other than 30°C (86°F) are encountered. When ambient temperatures other than 40°C (104°F) are encountered new **Table 310.15(B)(2)(b)** is then referenced. New numbered **Tables 310.15(B)(18)-(20)** also reference (footnotes) the use of new NEC section 310.15(B)(2) for ampacity correction factors when ambient temperature is other than 40°C (104°F).

Table 310.15(B)(2)(a) (Abbreviated) Ambient Temperature Correction Factors Based on 30°C (86°F)

Ambient Temperature (°C) Ⓐ	Temperature Rating of Conductor			Ambient Temperature (°F) Ⓑ
	60°C	75°C	90°C	
10 or less	1.29	1.20	1.15	50 or less
16-20	1.15 Ⓒ	1.11	1.08	60-68
26-30	1.00	1.00	1.00	78-86
46-50	0.58	0.75 Ⓒ	0.82	114-122
61-65	--	0.47	0.65	141-149
76-85	--	--	0.41 Ⓒ	168-176

For ambient temperatures other than 30°C (86°F), multiply the allowable ampacities specified in the ampacity tables by the appropriate correction factor shown below.

Table 310.15(B)(2)(a)

Ⓐ - Ambient Temperature °C (given in degrees Celsius)

Ⓑ - Ambient Temperature °F (given in degrees Fahrenheit)

C - Ampacity Correction Factors per Temperature Rating of Conductor (Used to adjust ampacity when ambient temperature is above or below 30°C (86°F)

PROBLEM-SOLVING METHODS BASED ON OPERATING CONDITIONS

Prior to the 2008 NEC there were only four methods per Article 310 for adjusting the ampacity of a conductor. With the addition of NEC (and Table) 310.15(B)(2)(c) to the 2008 NEC, conductors or cables enclosed in raceway that are exposed to sunlight on rooftops must also be considered for ampacity adjustments. However, just as the addition of NEC 310.15(B)(2)(c) seemed to sink in, the 2011 NEC incorporated numberical changes not only to this section [now 310.15(B)(3)(c)] but to other applicable ampacity adjustment sections along with introducing a new NEC 310.15(B)(2)(a), *Ambient Temperature Correction Factors.*

Now with the inclusion of NEC 310.15(B)(2)(a) along with NEC 310.15(B)(3)(c) where both sections will be discussed separately, the four methods for adjusting the ampacity of a conductor must be applied to safely determine the ampacity of a conductor. As a result, the operating conditions of any given conductor (regardless of size or insulation type; whether installed in raceway or cable jacket or directly buried) must be evaluated and deemed safe before placing the conductor in operation.

In order to develop a complete understanding of ampacity derating, each of these four methods will be thoroughly examined and respectively applied. However, for the purpose of familiarity, refer to the section, "Formulas for Adjusting Conductor Ampacity" before getting started.

Table 310.15(B)(16) - Ambient Temperature 30°C (86°F) - No More Than 3 Current-Carrying Conductors

METHOD 1

Operating Conditions
Ambient Temperature - **30°C (86°F)**
Number of Current Carrying Conductors - **3 or less**

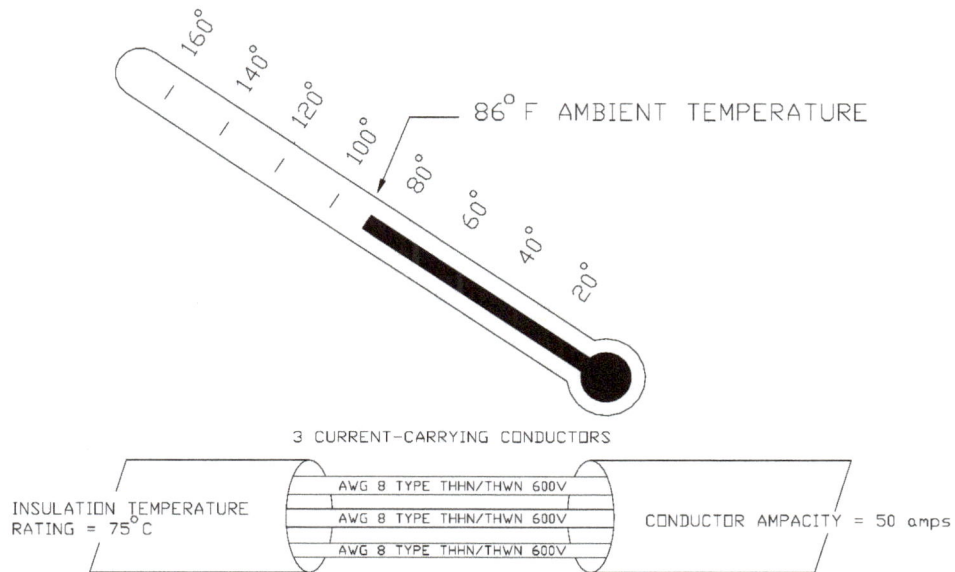

Figure 310.15(B)(16)-1

Discussion Question: Are the conductors shown in Figure 310.15(B)(16)-1 operating within the provisions of Table 310.15(B)(16)?

Answer: Yes. Because the conductors are operating within the provisions of Table 310.15(B)(16), that is, only three current-carrying conductors and installed in an area where the ambient temperature is 86°F (30°C), the ampacity of the conductors can be taken directly from Table 310.15(B)(16).

6. Determine the ampacities of the following conductors per Table 310.15(B)(16):

@ 60°C (140°F)
14 AWG TW copper   10 AWG RHW aluminum   8 AWG ZW-2 copper-clad aluminum
@ 75°C (167°F)
12 AWG ZW copper   6 AWG RHH aluminum   3 AWG THWN copper-clad aluminum
@ 90°C (194°F)
1 AWG MI copper   1/0 AWG THHN aluminum   350 kcmil SA copper-clad aluminum

@ 60°C (140°F)
14 AWG TW copper = 15 amps
10 AWG RHW aluminum = 25 amps
 8 AWG ZW-2 copper-clad aluminum = 35 amps

@ 75°C (167°F)
12 AWG ZW copper = 25 amps
 6 AWG RHH aluminum = 50 amps
 3 AWG THWN copper-clad aluminum = 75 amps

@ 90° C (194°F)
 1 AWG MI copper = 145 amps
1/0 AWG THHN aluminum = 135 amps
350 kcmil SA copper-clad aluminum = 280 amps

Table 310.15(B)(2)(a) - Correction Factors [Ambient Temperature exceeds 30°C (86°F)]
NEC 310.15(B)(3)(a) - No More Than Three Current-Carrying Conductors

METHOD 2

Operating Conditions

Ambient Temperature - **Greater than 30°C (86°F)**
Number of Current Carrying Conductors - **Three or less**

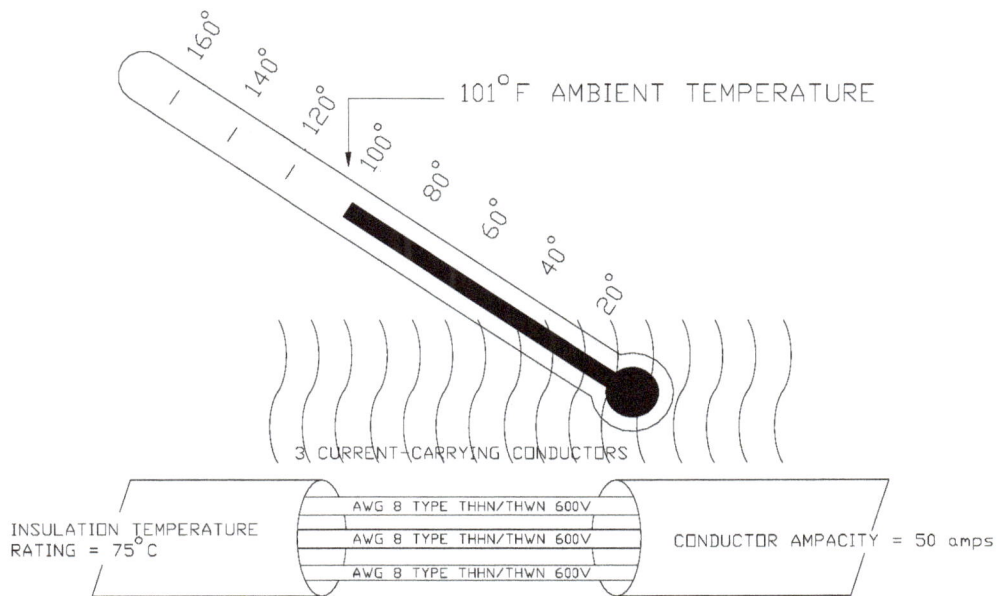

Figure 310.15(B)(16)-2

Discussion Question: Are the conductors shown in Figure 310.15(B)(16)-2 operating within the provisions of Table 310.15(B)(16)?

Answer: No. When a conductor is placed in an environment where the ambient temperature exceeds 30°C (86°F) the conductor will no longer operate within the limits of its insulation's temperature rating when operating at its allowable ampacity. If this occurs clearly one of three things must happen, **(1)** The ampacity of the conductor must be reduced to allow the conductor to operate safely within the limits of its insulation's temperature rating *or* **(2)** The conductor must be replaced with a conductor that's enclosed with insulation of a higher temperature rating *or* **(3)** The conductor must be of a larger size.

7. Three 8 AWG THWN (75°C) copper current-carrying conductors are installed in a raceway located in an ambient temperature of 101°F. Determine the ampacity of each conductor.

Conductor's ampacity per Table 310.15(B)(16)
8 AWG THWN Copper = 50 amps

FACTORS BASED ON ADVERSE CONDITIONS

Correction factors (CF) of Table 310.15(B)(2)(a) - @ 101°F CF = .88

Adjustment factors (AF) of Table 310.15(B)(3)(a)
(current-carrying conductors-ccc) - @ 3 ccc AF are not required

Conductor's allowable ampacity - 50 amps x .88 = 44 amps

Since only *three* current-carrying conductors are involved and they are installed in an area where the ambient temperature is above 86°F (30°C) the ampacity of each conductor is required to be reduced to ensure the temperature rating of the conductor's insulation is not exceeded. The conductor's new ampacity is 44 amps.

Table 310.15(B)(16) - Ambient Temperature 30°C (86°F)
NEC and **Table 310.15(B)(3)(a)** - More Than Three Current-Carrying Conductors – Adjustment Factors Required

METHOD 3

Operating Conditions

Ambient Temperature - **30°C (86°F)**
Number of Current Carrying Conductors - **5**

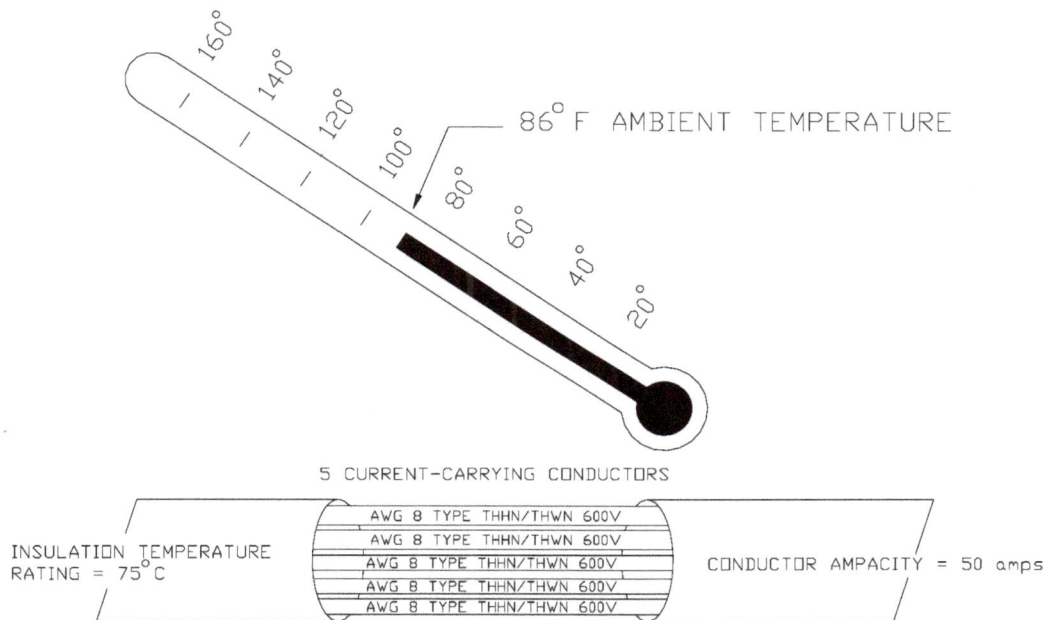

Figure 310.15(B)(16)-3

Discussion Question: Are the conductors shown in Figure 310.15(B)(16)-3 operating within the provisions of Table 310.15(B)(16)?

Answer: No. When more than 3 current-carrying conductors are installed in a raceway the conductors are exposed to more heat and less ventilation. The amount of heat produced by each conductor is proportional to the amount of current flowing through each conductor. If the amount of current is reduced so is the amount of heat. Therefore, **(1)** The ampacity of the conductor must be reduced to allow the conductor to operate safely within the limits of its insulation's temperature rating or **(2)** The conductor must be replaced with a conductor that's enclosed with insulation of a higher temperature rating or **(3)** The conductor must be of a larger size.

8. Five 8 AWG THWN (75°C) copper current-carrying conductors are installed in a raceway located in an ambient temperature of 86°F. Determine the ampacity of each conductor.

Conductor's ampacity per Table 310.15(B)(16)
8 AWG THWN Copper = 50 amps

FACTORS BASED ON ADVERSE CONDITIONS

Correction factors (CF) of Table 310.15(B)(2)(a) are not required

Adjustment factors (AF) of Table 310.15(B)(3)(a)
(current-carrying conductors-ccc) - @ 5 ccc AF = .80

Conductor's allowable ampacity - 50 amps x .80 = 40 amps

Since the conductors will be installed in an area where the ambient temperature is 86°F (30°C) the ampacity of each conductor is exactly as listed in Table 310.15(B)(16), 50 amps. However, the number of current-carrying conductors exceeds the limit; therefore the ampacity of each conductor is required to be reduced to ensure the temperature rating of the conductor's insulation is not exceeded. The conductor's new ampacity is now 40 amps.

Table 310.15(B)(16) - Correction Factors [Ambient Temperature exceeds 30°C (86°F)] **NEC** and **Table 310.15(B)(3)(a)** - More Than Three Current-Carrying Conductors – Adjustment Factors Required

METHOD 4

Operating Conditions

Ambient Temperature - **Greater than 30°C (86°F)**
Number of Current Carrying Conductors - **5**

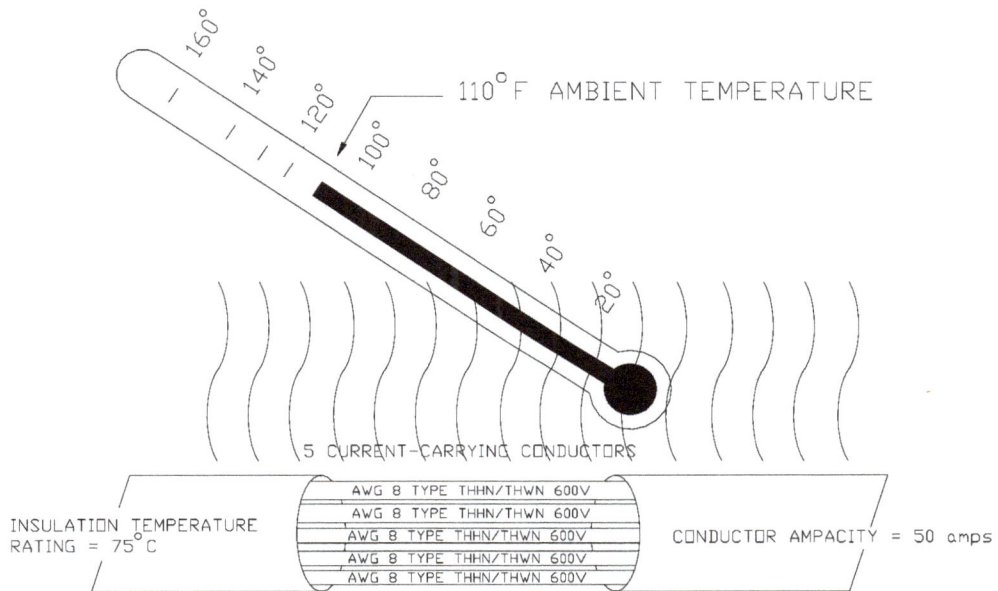

Figure 310.15(B)(16)-4

Discussion Question: Are the conductors shown in Figure 310.15(B)(16)-4 operating within the provisions of Table 310.15(B)(16)?

Answer: No. When more than three current-carrying conductors are installed in a raceway and the ambient temperature exceeds 86°F (30°C) the conductors are exposed to intensive heat, poor ventilation and incapable of dissipating heat at a normal rate. The amount of heat produced by each conductor is proportional to the amount of current flowing through each conductor combined with an ambient temperature that exceeds an unsafe limit. Again, if the insulation of the conductors is to operate within its temperature rating, the ampacity of the conductor must be reduced to allow the conductor to operate safely within the limits of its insulation's temperature rating or the conductor must be replaced with a conductor that's enclosed with insulation of a higher temperature rating or the conductor must be of a larger size.

9. Five 8 AWG THWN (75°C) copper current-carrying conductors are installed in a raceway located in an ambient temperature of 110°F. Determine the ampacity of each conductor.

Conductor's ampacity per Table 310.15(B)(16)
8 AWG THWN Copper = 50 amps

FACTORS BASED ON ADVERSE CONDITIONS

Correction factors (CF) of Table 310.15(B)(16) - @ 110°F CF = .82

Adjustment factors (AF) of Table 310.15(B)(3)(a)
(current-carrying conductors-ccc) - @ 5 ccc AF = .80

Conductor's allowable ampacity - 50 amps x .82 x .80 = 32.8 amps

Since conductors will be installed in an area where the ambient temperature exceeds 86°F (30°C) the ampacity of each conductor is required to be reduced. Furthermore, since the number of current-carrying conductors exceeds the limit, the ampacity of each conductor is required to be reduced even more. The conductor's new ampacity is now 32.8 amps and the insulation of these conductors can now withstand these adverse conditions.

Examples Based On The Conditions Of **METHOD 1** (10. - 11.) [2]

10. What is the allowable ampacity of two 1/0 AWG THHN copper conductors installed in an area with a surrounding ambient temperature of 86°F?

Conductor's ampacity per Table 310.15(B)(16)
1/0 AWG THHN Copper = 170 amps

FACTORS BASED ON ADVERSE CONDITIONS

Correction factors (CF) of Table 310.15(B)(2)(a) are not required. The ambient temperature is within limits.

Adjustment factors (AF) of Table 310.15(B)(3)(a) - (current-carrying conductors-ccc) are not required. The number of current-carrying conductors is less than three.

No adjustments are required, the ampacity of the conductors is the same as listed in Table 310.15(B)(16), 170 amps.

11. Six 1 AWG THW aluminum conductors pass through a 20-inch raceway to enter a meter can. The ambient temperature is 30°C. What is the allowable ampacity of the conductors?

Conductor's ampacity per Table 310.15(B)(16)
1 AWG THW Aluminum = 100 amps

FACTORS BASED ON ADVERSE CONDITIONS

Correction factors (CF) of Table 310.15(B)(2)(a) are not required. The ambient temperature is within limits.

Adjustment factors (AF) of Table 310.15(B)(3)(a) - (current-carrying conductor-ccc) are not required. Although the number of current-carrying conductors exceeds three, according to NEC 310.15(B)(3)(a)(2) adjustments are not required because the raceway (nipple) which the conductors will pass through is less than 24 inches in length.

Therefore, no adjustments are required, the ampacity of the conductors is the same as listed in Table 310.15(B)(16), 100 amps.

Examples Based On The Conditions Of **METHOD 2** (12. - 13.) [2]

12. What is the allowable ampacity of three 4 AWG THW copper conductors located in an area where the ambient temperature is 118°F?

Conductor's ampacity per Table 310.15(B)(16)
4 AWG THW Copper = 85 amps

FACTORS BASED ON ADVERSE CONDITIONS

Correction factors (CF) of Table 310.15(B)(2)(a) - @ 118°F CF = .75

Adjustment factors (AF) of Table 310.15(B)(3)(a)
(current-carrying conductors-ccc) - @ 3 ccc AF are not required

Conductor's allowable ampacity - 85 amps x .75 = 63.75 amps

13. The ambient temperature in a laundry facility is 97°F. Three individual circuits are installed in separate conduit passing through the facility. What is the allowable ampacity of each conductor based on the following information?

CONDUIT 1	**CONDUIT 2**	**CONDUIT 3**
3-10 AWG Al THWN	3-2 AWG Cu TW	3-8 AWG Al THW-2

CONDUIT 1

Conductor's ampacity per Table 310.15(B)(16)
10 AWG THWN Aluminum = 30 amps

FACTORS BASED ON ADVERSE CONDITIONS

Correction factors (CF) of Table 310.15(B)(2)(a) - @ 97°F CF = .88

Adjustment factors (AF) of Table 310.15(B)(3)(a)
(current-carrying conductors-ccc) - @ 3 ccc AF are not required

Conductor's allowable ampacity - 30 amps x .88 = 26.4 amps

CONDUIT 2

Conductor's ampacity per Table 310.15(B)(16)
2 AWG TW Copper = 95 amps

FACTORS BASED ON ADVERSE CONDITIONS

Correction factors (CF) of Table 310.15(B)(2)(a) - @ 97°F CF = .82

Adjustment factors (AF) of Table 310.15(B)(3)(a)
(current-carrying conductors-ccc) - @ 3 ccc AF are not required

Conductor's allowable ampacity - 95 amps x .82 = 77.9 amps

CONDUIT 3

Conductor's ampacity per Table 310.15(B)(16)
8 AWG THW-2 Aluminum = 45 amps

FACTORS BASED ON ADVERSE CONDITIONS

Correction factors (CF) of Table 310.15(B)(2)(a) - @ 97°F CF = .91

Adjustment factors (AF) of Table 310.15(B)(3)(a)
(current-carrying conductors-ccc) - @ three ccc AF are not required

Conductor's allowable ampacity - 45 amps x .91 = 40.95 amps

Examples Based On The Conditions Of **METHOD 3** (14. - 15.) [2]

14. Two 500 kcmil XHHW-2 copper conductors are installed in parallel per phase with neutrals to feed a 4W, 3φ, 480/277V system consisting primarily of nonlinear loads. Since the conductors (8) will be installed in the same conduit, what is the allowable ampacity of the conductors?

nonlinear load - Equipment which operates on the principles of inproportionality. In *nonlinear loads* the operating current and voltage are inproportional to one another. The operating current in this type of load neither increases nor decreases in proportion to the operating voltage. These loads produce voltage and sine waves that are non-sinusoidal. Equipment such as computers, converters, data-processing, drives (adjustable/frequency/speed/variable), electronic ballasts, electric discharge lighting (fluorescent, high and low-pressure sodium, mercury-vapor, metal-halide, etc.), inverters, medical and laboratory test equipment, programmable logic controllers (PLC), UPS systems, welders, etc. are *nonlinear loads.*

Conductor's ampacity per Table 310.15(B)(16)
500 kcmil XHHW-2 Copper = 430 amps

FACTORS BASED ON ADVERSE CONDITIONS

Correction factors (CF) of Table 310.15(B)(2)(a) are not required. The ambient temperature is assumed to be 30°C (86°F) since an actual temperature is not given.

Adjustment factors (AF) of Table 310.15(B)(3)(a) - (current-carrying conductors-ccc) @ 8 ccc AF = .70. Since the major portions of the loads are nonlinear, all enclosed conductors must be considered current-carrying per NEC 310.15(B)(5)(c).

Conductor's allowable ampacity - 430 amps x .70 = 301 amps

15. Assume the loads in question No. 14. are primarily linear loads.

linear load - Equipment which operates on the principles of proportionality. In *linear loads* the operating current and voltage are proportional to one another. As the load's operating current increases or decreases the operating voltage increases or decreases respectively. These loads produce voltage and sine waves that are sinusoidal. Electric motors, heating equipment, resistive lighting (incandescent), etc. are *linear loads.*

Conductor's ampacity per Table 310.15(B)(16)
500 kcmil XHHW-2 Copper = 430 amps

FACTORS BASED ON ADVERSE CONDITIONS

Correction factors (CF) of Table 310.15(B)(16) are not required. Again, the ambient temperature is assumed to be 30°C (86°F) since an actual temperature is not given.

Adjustment factors (AF) of Table 310.15(B)(3)(a) - (current-carrying conductors-ccc) @ 6 ccc AF = .80. Since the major portions of the loads are linear, the neutral conductors are considered non-current carrying and therefore not counted per NEC 310.15(B)(5)(c).

Conductor's allowable ampacity - 430 amps x .80 = 344 amps

Examples Based On The Conditions Of **METHOD 4** (16. - 21.) [6]

16. There are eleven 12 AWG THW copper conductors installed in the same raceway. Of the 11 copper conductors 10 are current-carrying. The surrounding temperature in the area is approximately 73°F. What is the allowable ampacity of the conductors?

Conductor's ampacity per Table 310.15(B)(16)
12 AWG THW Copper = 25 amps

FACTORS BASED ON ADVERSE CONDITIONS

Correction Factors (CF) of Table 310.15(B)(2)(a) - @ 73°F CF = 1.05. Although the ambient temperature is less than 86°F, the Correction Factors of Table 310.15(B)(2)(a) are still applied.

Adjustment factors (AF) of Table 310.15(B)(3)(a) - (current-carrying conductors-ccc) @ 10 ccc AF = .50

Conductor's allowable ampacity - 25 amps x 1.05 x .50 = 13.13 amps

17. Five 1 AWG THW aluminum current-carrying conductors are enclosed in the same raceway. If the raceway is installed in an area where the ambient temperature is 48°C, determine the allowable ampacity of the conductors.

Conductor's ampacity per Table 310.15(B)(16)
1 AWG THW Aluminum = 100 amps

FACTORS BASED ON ADVERSE CONDITIONS

Correction factors (CF) of Table 310.15(B)(2)(a) - @ 48°C CF = .75

Adjustment factors (AF) of Table 310.15(B)(3)(a) - (current-carrying conductors-ccc) @ 5 ccc AF = .80

Conductor's allowable ampacity - 100 amps x .80 x .75 = 60 amps

18. An electrical metallic tubing containing four 8 AWG THW copper conductors and five 10 AWG THWN copper conductors are routed through an area where the ambient temperature is 88°F. If the enclosed conductors are current-carrying, determine the allowable ampacity of each set of conductors.

Conductor's ampacity per Table 310.15(B)(16)
10 AWG THWN copper = 35 amps
 8 AWG THW copper   = 50 amps

FACTORS BASED ON ADVERSE CONDITIONS

Correction factors (CF) of Table 310.15(B)(2)(a) - @ 88°F CF = .94

Adjustment factors (AF) of Table 310.15(B)(3)(a) - (current-carrying conductor-ccc) @ 9 ccc AF = .70

Conductor's allowable ampacity

10 AWG - 35 amps x .70 x .94 = 23.03 amps
8 AWG - 50 amps x .70 x .94 = 32.9 amps

19. A raceway containing nine 10 AWG THWN copper (current-carrying) conductors are run in an area where the ambient temperature is 105°F. Three of the conductors supply a three-phase, 480 volt, 15hp motor. Determine the ampacity of the motor's conductors. If the conductors cannot be used, what size is required?

The ampacity of the conductors has to be adjusted to allow for adverse conditions. According to the Correction Factors of Table 310.15(B)(2)(a) and the Adjustment Factors of Table 310.15(B)(3)(a), the conductor's ampacity as listed in Table 310.15(B)(16) must be reduced by .82 and .70 respectively.

35 amps x .82 x .70 = 20.09 amps

The ampacity of the conductors is limited to a bit over 20 amps. Because the full load current of the motor alone is 21 amps (Table 430.250) it's clear that the 10 AWG conductors cannot be used. If the adverse conditions did not exist they could be used. Remembering the requirements of NEC 430.22 and the stipulations based on the adverse conditions, the correct size conductors can be determined per calculated ampacity.

$$\text{Required conductor ampacity} = \frac{21A \times 1.25}{.82 \times .70} = 45.73A$$

Based on the results of the calculation, 8 AWG THWN copper conductors which have an ampacity of 50 amps are required to supply the motor.

20. A two-wire 277V circuit and a three-wire 480V three-phase feeder are installed in the same conduit. The feeder will supply a 59 amps continuous load and a 65 amps noncontinuous load. Primarily, the circuits will be exposed to a surrounding temperature of 34°C. If the feeder circuit is protected by a circuit breaker with an unmarked terminal rating, what size THW-2 copper conductors can be used to supply the loads?

Before getting started refer to the section, "Formulas for Adjusting Conductor Ampacity" of this Article.

When feeder conductors supply a combination of continuous and noncontinuous loads, NEC 215.2(A)(1) states that the minimum feeder circuit conductor size, before the application of any adjustment or correction factors must have an allowable ampacity not less than the noncontinuous load plus 125 percent of the continuous load.

Although the THW-2 copper conductors are rated for 90°C, the ampacity of the conductors must be determined at 75°C as required by NEC 110.14(C)(1)(b)(2).

Per Table 310.15(B)(2)(a), at 34°C, the THW-2 copper conductors will have a .94 correction factor at 75°C and a .80 adjustment factor per Table 310.15(B)(3)(a) based on five current-carrying conductors. When calculated,

$$\frac{59A \times 1.25 + 65A}{.94 \times .80} = 184.51A$$

Based on the calculated ampacity (184.51A), 3/0 AWG THW-2 conductors are required. The conductors have a rated ampacity of 225 amps at 90°C yet limited to 200 amps at 75°C.

21. Three runs of 10/3 ZW cable in lengths of 15 feet are being installed in direct contact with each other in the ceiling of a machine shop. During the summer months the shop has been known to reach temperatures up to 115°F. At what ampacity must the cables be adjusted to operate within their limits?

Conductor's ampacity per Table 310.15(B)(16)
10 AWG ZW copper = 35 amps

FACTORS BASED ON ADVERSE CONDITIONS

Correction factors (CF) of Table 310.15(B)(16) - @ 115°F CF = .75

Table 310.15(B)(16) shows the temperature rating for this type insulation to be 75°C. However, Table 310.104(A) reflects various maximum operating temperatures based on ZW type insulation. In wet locations the temperature rating is limited to 75°C, in dry locations the temperature rating is limited to 90°C and in special applications (dry) the temperature rating is limited to 150°C. Because it is safe to assume that the area above the ceiling is a dry location and the type loads these conductors will serve are unknown, the insulation will be based on a temperature rating of 90°C. The correction factor should be based on 90°C.

Conductor's ampacity per Table 310.15(B)(16)
10 AWG ZW copper (90°C) = 40 amps

Correction factors (CF) of Table 310.15(B)(2)(a) - @ 115°F CF = .82

NEC 310.15(B)(3)(a) states that where single conductors or multiconductor cables are continuously longer than 24 inches without maintaining spacing and are not installed in raceways, the allowable ampacity of each conductor shall be reduced per Table 310.15(B)(3)(a). Since each cable will be installed in direct contact with each other exceeding 24 inches (15 feet), the adjustment factors must be applied.

Adjustment factors (AF) of Table 310.15(B)(3)(a) (current-carrying conductors-ccc) - @ 9 ccc AF = .70

Conductor's allowable ampacity

10 AWG - 40 amps x .82 x .70 = 22.96 amps

310.15(B)(2) - Ambient Temperature Correction Factors

22. Using the equation provided in NEC 310.15(B)(2), determine the permitted ampacity of a 10 AWG THW copper conductor if exposed to an ambient temperature of 41°C. Also determine such permitted ampacities for a 4 AWG THWN copper conductor and a 2/0 AWG THW aluminum conductor if the conductors are exposed to 93°F and 20°C respectively.

The addition of this new section provides an alternate means to calculate a specific ampacity of a conductor (when exposed to an ambient temperature other than 30°C or 86°F) opposed to using an exclusive correction factor for a range of ambient temperatures as new Table 310.15(B)(2)(a) still does in likeness to the previously used Correction Factors of Table 310.16. Before the existence of this new section for example, if the corrected ampacity of the 10 AWG THW copper conductor (which has a temperature rating of 75°C) was requested at 41°C, the same results would be realized if the ambient temperature was 42°C, 43°C, 44°C or 45°C. This is so because the same correction factor per individual ambient temperature according to Table 310.15(B)(2)(a) would still be applied, .82.

With the new equation, a specific correction factor can be determined based on a given ambient temperature which now permits a corrected conductor ampacity to be calculated that corresponds to the derived correction factor.

Before getting started let's clear up a few things. Underneath the equation in NEC 310.15(B)(2) two unfamiliar references are listed, I' and T_a'. The acute accent symbol (′) is used to distinguish one reference from another. Compared to an ampacity listed in a table which is commonly identified by the capital letter I, I' identifies a corrected ampacity based on an ambient temperature other than the norm, 30°C (86°F) or 40°C (104°F). On the other hand, T_a references the insertion of either temperature value, 30°C (86°F) or 40°C (104°F while T_a' identifies the use of an ambient temperature other than the norm. In mathematical terms, the reference I' is as expressed as "**I prime**" while T_a' is expressed as, "**T sub a prime**" where the symbol (′) is used instead to identify the term, **prime.** Again, the term **prime** in this situation refers to the difference between the norm and a specific ambient temperature whether above or below the norm. Now observe.

In this question T_a' is 41°C, an ambient temperature other than the norm. Applying the equation,

$$I' = I \sqrt{\frac{T_c - T_á}{T_c - T_a}}$$

where the ampacity per Table 310.15(B)(16) = 35A at 30°C results to,

$$I' = 35A \times \left(\sqrt{\frac{75 - 41}{75 - 30}} = \sqrt{\frac{34}{45}} = \sqrt{.756} = .869 \right)$$

where, .869 represents the value that will cause the ampacity of the 10 AWG THW copper conductor to be reduced based on an ambient temperature that is above the standard ambient temperature. As a result, the corrected ampacity based on the given ambient temperature is,

$$I' = 35A \times .869 = 30.42A$$

If the correction factor given in Table 310.15(B)(2)(a) was applied (based on the given ambient temperature) an approximate corrected ampacity would be derive instead where,

$$35A \times .82 = 28.7A$$

This approximate corrected ampacity would be the same if the given ambient temperature was 42°C, 43°C, 44°C or 45°C based on the correction factor of Table 310.15(B)(2)(a). Again, with this new section and equation a specific corrected ampacity can be determined.

4 AWG THWN copper conductor at 93°F

Ampacity per Table 310.15(B)(16) = 85A at 167°F

$$I' = 85A \times \left(\sqrt{\frac{167-93}{167-86}} = \sqrt{\frac{74}{81}} = \sqrt{.914} = .956 \right)$$

$$I' = 85A \times .956 = 81.26A$$

If the correction factor given in Table 310.15(B)(2)(a) was applied (based on the given ambient temperature) an approximate corrected ampacity would be derived instead where,

$$85A \times .94 = 79.9A$$

2/0 AWG aluminum THW conductor at 20°C

Ampacity per Table 310.15(B)(16) = 135A at 75°C

$$I' = 135A \times \left(\sqrt{\frac{75-20}{75-30}} = \sqrt{\frac{55}{45}} = \sqrt{1.22} = 1.105 \right)$$

$$I' = 135A \times 1.105 = 149.18A$$

If the correction factor given in Table 310.15(B)(2)(a) was applied (based on the given ambient temperature) an approximate corrected ampacity would be derived instead where,

$$135A \times 1.11 = 149.85A$$

Because the derived correction factor (1.105) and the given correction factor (1.11) are approximately the same, both corrected ampacities yield approximate values.

310.15(B)(3)(a) - Adjustment Factors (Conductors of Different Systems)

23. Six - 14 AWG, 6 - 8 AWG, and 3 - 6 AWG THW copper conductors are installed in the same raceway.

    The conductors are used for the following purposes:

        3 - 14 AWG - Signal circuits   3 - 14 AWG - Control circuits
        6 - 8 AWG - Lighting circuits   3 - 6 AWG - Motor circuits

What are the allowable ampacities of each conductor?

The total number of current-carrying conductors in the raceway is 15. Based on 15 current-carrying conductors per Table 310.15(B)(3)(a), the adjusted factor is .50.

However, according to NEC 310.15(B)(3)(a), only the lighting and power conductors are required to be derated, the signal and control circuit conductors are not. Therefore, the use of the adjustment factor only applies to the 9 lighting and motor circuit conductors instead of all 15 enclosed conductors. The adjustment factor in this case can now be increased to .70.

At 75°C, the ampacity of the 8 AWG copper conductors is 50 amps and the ampacity of the 6 AWG copper conductors is 65 amps. The ampacity of both conductors is now multiplied by the .70 adjusted factor and the allowable ampacities of these conductors are as calculated.

(8 AWG) 50 amps x .70 = 35 amps
(6 AWG) 65 amps x .70 = 45.5 amps

As far the 14 AWG signal and control circuit conductors, the ampacity of these conductors can be taken directly from Table 310.15(B)(16) which is 20 amps.

Suppose three additional 8 AWG conductors were installed in the raceway for future use, would this cause further ampacity adjustments?

The heading of the first column of Table 310.15(B)(3)(a) in the 2011 NEC now reads with a referenced footnote, "Number of Conductors" opposed to previous editions which read, "Number of Current-Carrying Conductors." The footnote states that the number of conductors is the total number of conductors in the raceway or cable adjusted in accordance with 310.15(B)(5) and (6). As a result, future or spare conductors must be counted because of the possibilities of being used at some later time and become "current-carrying". Therefore, the three 8 AWG conductors must be counted as "current-carrying" thus requiring the adjustment factor to be reduced to 50 percent (.50) based on twelve (12) current-carrying conductors. At 50 percent, further ampacity adjustments are required resulting to,

(8 AWG) 50 amps x .50 = 25 amps
(6 AWG) 65 amps x .50 = 32.5 amps

310.15(B)(3)(a) Informational Note No. 1 - Adjustment Factors (More Than Three Current-Carrying Conductors in Raceway or Cable with Load Diversity) (24. - 25.) [2]

24. Determine the allowable ampacity of 12 - 6 AWG XHHW copper current-carrying conductors that are enclosed in electrical metallic tubing along with 18 other conductors.

Based on the given question, the allowable ampacity of the given conductors can be determined in the same manner as those previously calculated [applying the provisions of Table 310.15(B)(3)(a)] since there is no load diversity to be considered.

25. Now determine the allowable ampacity of the same conductors in question No. 24. where the 18 other conductors may become current-carrying at different times.

NEC 310.15(B)(3)(a) Informational Note No. 1 reference the use of Table B.310.15(B)(2)(11) of Informative Annex B. However, before proceeding it should be understood that all information found in Informative Annex B is for informational purposes only and therefore is not code enforceable. According to NEC B.310.15(B)(1), the intended purpose of Informative Annex B is to provide application information for ampacities calculated under engineering supervision.

Referring now to Table B.310.15(B)(2)(11), a 60 percent adjustment factor must be applied when 25-42 current-carrying conductors are enclosed in raceway. This adjustment factor is based on the enclosed current-carrying conductors operating at a 50 percent load diversity.

Before answering the question it is best to understand the term *load diversity*. Depending on the application, there could exist an operating condition where conductors supplying a certain load or loads may not be energized simultaneously (at the same time). Such operating conditions define the term *load diversity*. When a 50 percent load diversity occurs only half of the conductors in a cable or the conductors enclosed in a raceway will operate at the same time. The adjustment factors listed in Table B.310.15(B)(2)(11) are based on operating conditions where a 50 percent load diversity may or may not exist.

According to Table 310.15(B)(16), a 6 AWG XHHW copper conductor has an ampacity of 65 amps. Table B.310.15(B)(2)(11) provides a formula and related information for calculating the allowable ampacity of current-carrying conductors in raceway or cable when a 50 percent load diversity exist. Using the formula,

$$A_2 = \left(\sqrt{\frac{0.5N}{E}} \times (A_T \times AF) \right) \text{ or } A_1, \text{ whichever is less}$$

where A_2 represents the ampacity limit for the current-carrying conductors in the raceway (or cable)

Applying the formulas where,

N = 18 [total number of conductors used to select adjustment factor from Table B.310.15(B)(2)(11)]

A₁ = Ampacity **(A_T)** from Tables 310.15(B)(16) and (18), Tables B.310.15(B)(2)(1), (6) or (7) multiplied by the appropriate adjustment factor from Table B.310.15(B)(2)(11)

AF = Adjustment factor from Table B.310.15(B)(2)(11)

E = Number of conductors carrying current simultaneously in raceway or cable

results to,

$$A_2 = \sqrt{\frac{0.5 \times 18}{12}} \times (65A \times .70)$$

$$= .866 \times (65A \times .70) = 39.4A$$

Considering the 50 percent load diversity, the ampacity of the 6 AWG XHHW copper conductors is limited to A_2 (39.4A) which is less than A_1 (45.5A [65A x .70]).

310.15(B)(3)(a)(1) - Adjustment Factors (Conductors in Cable Tray)

See question Nos. 1. and 2. of Section 392.22(A)(1)(a).

310.15(B)(3)(a)(2) - Adjustment Factors (Nipples)

26. Twenty-seven 12 AWG THW aluminum conductors are installed in a 1½" raceway that's eighteen inches long. The normal temperature where these conductors will be installed is usually 123°F. Determine the allowable ampacity of these conductors.

NEC 310.15(B)(3)(a)(2) states that the adjustment factors shall not apply to conductors in raceway having a length not exceeding 24" (nipple - measuring 24" or less).

Therefore, because the 1½" raceway enclosing the conductors is only 18" in length and classified as a nipple, the 27 aluminum conductors are only required to be derated based on the adverse ambient temperature. As a result, in applying the correction factor of Table 310.15(B)(2)(a), [.67 at 123°F] and the conductor's ampacity per Table 310.15(B)(16) [20A at 75°C] the allowable ampacity of the conductors is, 13.4A [20A x .67].

310.15(B)(3)(a)(3) - Adjustment Factors (Underground Conductors)

See question No. 36.

310.15(B)(3)(c) - Circular Raceways Exposed to Sunlight on Rooftops (Adjustment Factors)

27. Two 4-wire 250 kcmil THWN copper feeder conductors are installed in 3" EMT. The EMT is installed 4½" above the roof of a building that's directly exposed to the sun. If the building is located in Shreveport, Louisiana, calculate the ampacity of the feeder conductors if six conductors are current carrying.

NEC 310.15(B)(3)(c) states, *Where conductors or cables are installed in circular raceways exposed to direct sunlight on or above rooftops, the adjustments shown in Table 310.15(B)(3)(c) shall be added to the outdoor temperature to determine the applicable ambient temperature for application of the correction factors in Table 310.15(B)(2)(a) or Table 310.15(B)(2)(b).*

Based on the requirements of NEC 310.15(B)(3)(c), in order to make such adjustments an outdoor temperature (of a specific location) must be known to determine the applicable ambient temperature for application of the correction factors per given tables.

NEC 310.15(B)(3)(c) Informational Note provides a reference for obtaining the average ambient temperatures in various locations based on the ASHRAE (American Society of Heating, Refrigerating and Air-Conditioning Engineers, Inc.) Handbook – Fundamentals. In conjunction with the data provided by the ASHRAE, the Copper Development Association Inc. (CDA), an organization which caters to the copper industry, developed a revised table listing the temperature inside conduit where exposed to direct sunlight for various US cities. In the table a two (2) percent ASHRAE design temperature is provided for each listed city which resulted from averaged temperatures covering the months of June through August. In addition, the table also lists the likely temperatures inside unfilled rooftop conduit when exposed to direct sunlight where conduit is installed ½" up to 36" above a roof.

Based on such gathered data and the need to address the concern of conduits exposed to sunlight on rooftops, Table 310.15(B)(3)(c) was introduced into the 2008 NEC. Whether using the pre-determined data of the CDA table or the adjustment requirements of Table 310.15(B)(3)(c) where the average (outdoor) ambient temperature is required for both tables, the results of either are synonymous.

For the city of Shreveport, Louisiana the CDA Table list a 2 percent design temperature or an average ambient temperature of 98°F. Since the question states that the EMT is installed 4½" above the roof per Table 310.15(B)(3)(c), a 30°F temperature adder is applied with the average ambient temperature which results to 128°F. In comparison to the data provided in the CDA table for the city of Shreveport, Louisiana the results are exactly the same.

Because 250 kcmil THWN copper conductors are being used, the correction factors of Table 310.15(B)(2)(a) is applied. At 128°F, a .67 correction factor must be applied along with an .80 adjustment factor per Table 310.15(B)(3)(a) to the 255A rated ampacity of a 250 kcmil THWN copper conductor. As a result, the ampacity of the feeder conductors must be limited to 136.68A, where 255A x .67 x .80 = 136.68A.

Prior to the 2008 National Electrical Code, NEC 310.10, fine print note No. 2 of the 2005 National Electrical Code gave reference to a temperature rise above the ambient temperature for conductors installed in conduit exposed to direct sunlight in close proximity to rooftops.

310.15(B)(4) - Bare or Covered Conductors

Where bare or covered conductors are installed with insulated conductors, the temperature rating of the bare or covered conductor shall be equal to the lowest temperature rating of the insulated conductors for the purpose of determining ampacity.

THE NEUTRAL CONDUCTOR

310.15(B)(5) - Neutral Conductor

NEC 310.15(B)(5) provides three (3) conditions for determining whether neutral conductors are classified and counted as current-carrying conductors. However, before proceeding we need to define what a neutral conductor is and identify its application in an electrical system.

According to the definition provided in Article 100, a *neutral conductor* is the conductor connected to the "neutral point" of a system that is intended to carry current under normal conditions. To fully grasp the definition, the inclusive term "neutral point" is also defined in Article 100 as "The common point on a wye-connection in a polyphase (two or three phase) system or midpoint on a single-phase, 3-wire system, or midpoint of a single-phase portion of a 3-phase delta system, or a midpoint of a 3-wire, direct-current system."

With such valued explanation, the definition of a *neutral conductor* can now be re-written as, "the conductor connected to the **(a)** common point on a wye-connection in a polyphase system *or* **(b)** midpoint on a single-phase, 3-wire system, *or* **(c)** midpoint of a single-phase portion of a 3-phase delta system, *or* **(d)** a midpoint of a 3-wire, direct-current system; that is intended to carry current under normal conditions". For further understanding of a now clearer definition, refer to the following illustrations:

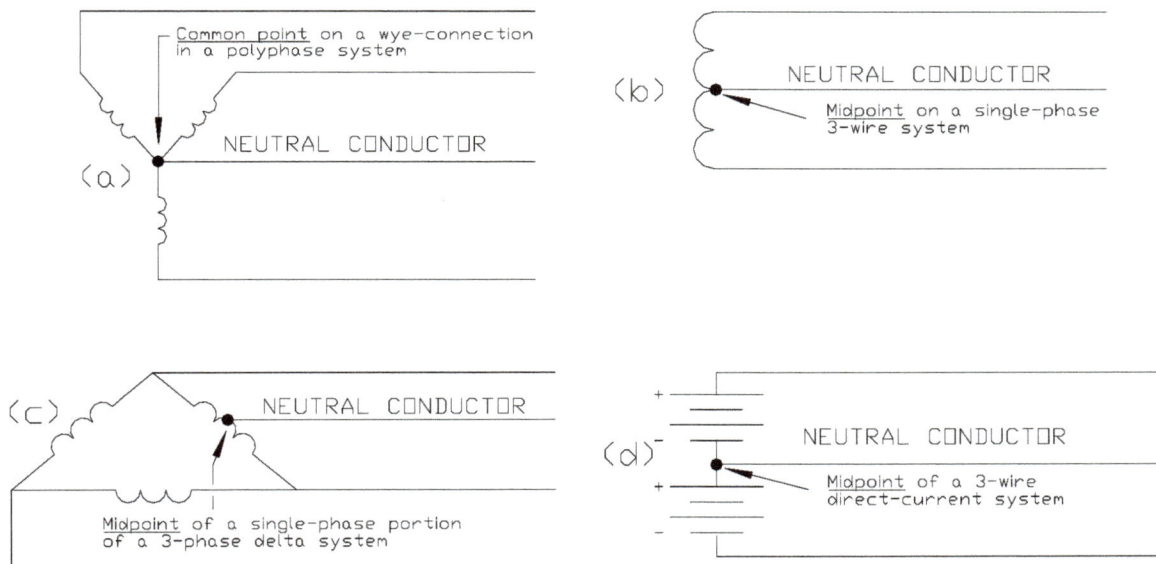

Based on the NEC definition of a neutral conductor the following applications apply per labeled illustration with the exception of illustration **(d)**:

(a) a 3φ-208/120V or 480/277V-4W circuit*
a 3φ-208/120V or 480/277V-4W system*
(b) a 1φ-240/120V, 3W circuit
a 1φ-240/120V, 3W system
(c) a 1φ-240/120V, 3W system originating from a 3φ-4W-Delta (Δ) source

With better knowledge of what a *neutral conductor* is, other type applications that are thought to include or involve a neutral conductor are identified as follows for what they actually are or recognized as in the NEC:

> **(1)** 2W-120V or 277V circuit (~~neutral~~ grounded conductor)
>
> **(2)** 1φ-3W-208/120V circuit *or* 1φ-3W-208/120V system originating from a 3φ-208/120V source* (~~neutral~~ common conductor)

*When consisting only of linear (resistive or non-inductive) loads harmonic currents are not present.

Having now clarified the definition and the various applications of a neutral conductor we can now consider the provisions of NEC 310.15(B)(5) and how each provision relates per given application.

310.15(B)(5)(a)

A *neutral conductor* that carries only the unbalanced current from other conductors of the same circuit need not be counted when applying the provisions of NEC 310.15(B)(3)(a).

In this case NEC 310.15(B)(5)(a) reference 3W-240/120V-1φ systems (circuits) [whether dedicated or originating from a 4W-240/120V-3φ Delta (Δ) connected system] and 4W-208/120V or 480/277-3φ wye (Y) connected systems (circuits) that supply linear loads where harmonic currents are not present. Refer to Figures 310.15(B)(5)(a).1 - .3

NEUTRAL CONDUCTOR NOT COUNTED as "current-carrying"

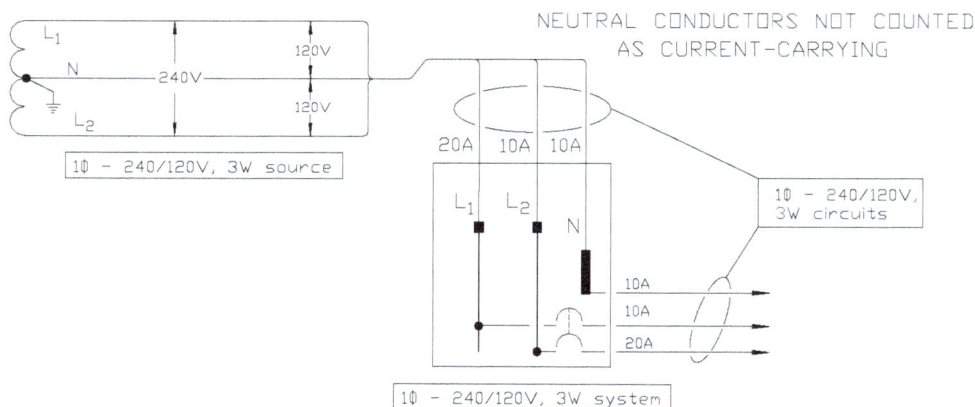

Figure 310.15(B)(5)(a).1

Figure 310.15(B)(5)(a).2

$$I_N = \sqrt{(15A)^2 + (12A)^2 + (9A)^2 - (15A \times 12A) + (12A \times 9A) + (9A \times 15A)} = 5.2A$$

Figure 310.15(B)(5)(a).3

In each of the above situations the neutral conductors will only carry the unbalanced current of the neutral loads or theoretically "0" amps if the neutral loads are balanced. Therefore, the neutral conductors are considered non-current carrying and are not counted. Figure 310.15(B)(5)(a).3 reflects three individual circuits sharing a common neutral.

Now observe the 2W circuits in Figures 310.15(B)(5)(a).4 and .5. Regardless of the amount of current flowing through the ungrounded (hot) conductors the same amount of current will flow through the neutral or in this situation the grounded conductors. Remember, neutral conductors are mostly recognized for carrying only unbalanced current although the terms *grounded* or *neutral* are often used interchangeably depending upon the application. Now with a 2W circuit the grounded conductor does not carry an unbalanced current but rather all of the circuit's current and is therefore classified as a current-carrying conductor. As a result, the grounded conductor must be counted as a current-carrying conductor and the provisions of NEC 310.15(B)(3)(a) are applied.

GROUNDED CONDUCTOR COUNTED as "current-carrying"

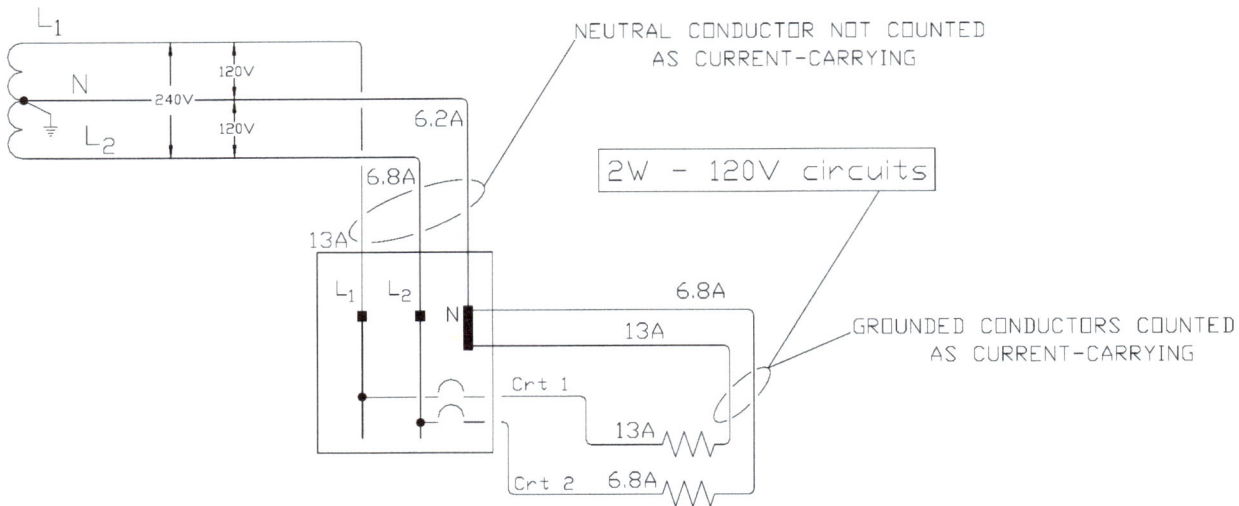

Figure 310.15(B)(5)(a).4

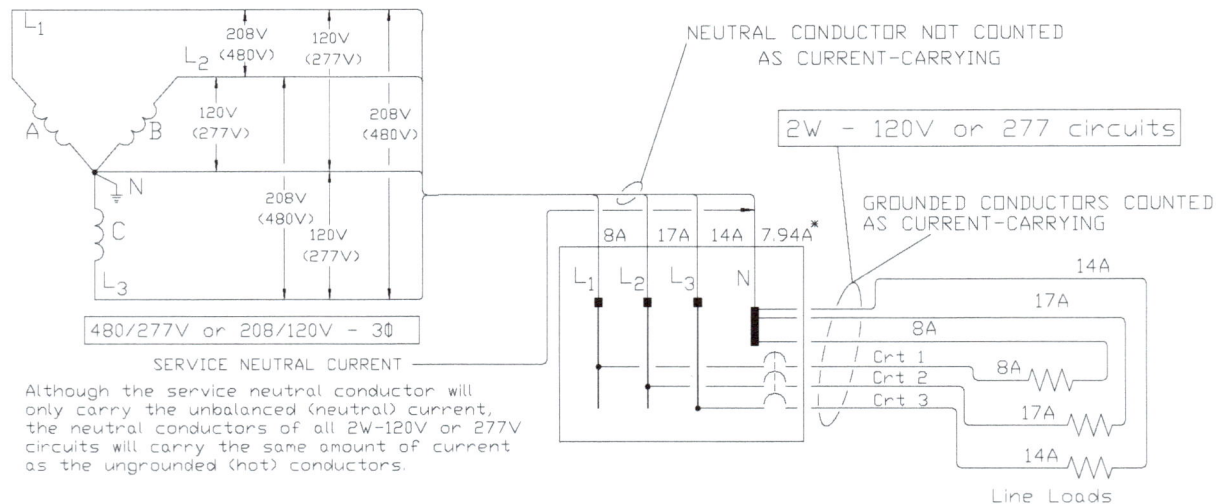

Figure 310.15(B)(5)(a).5

$$I_N = \sqrt{(8A)^2 + (17A)^2 + (14A)^2 - (8A \times 17A) + (17A \times 14A) + (14A \times 8A)} = 7.94A$$

In any case, if the two branch circuits in Figure 310.15(B)(4)(a).4 or at least two of the branch circuits in Figure 310.15(B)(5)(a).5 were installed in the same raceway the ampacity of the conductors would require derating per NEC 310.15(B)(3)(a).

310.15(B)(5)(b)

In a 3-wire circuit consisting of two phase (ungrounded) conductors and the *neutral conductor* of a 4-wire, 3 phase wye (Y) connected system, a common conductor carries approximately the same current as the line-to-neutral load currents of the other conductors and shall be counted when applying the provisions of NEC 310.15(B)(3)(a).

In this case NEC 310.15(B)(5)(b) reference 3W, 1φ, 208/120V circuits that are supplied from 4W, 3φ, 208/120V Wye (Y) connected systems. Refer to Figure 310.15(B)(5)(b).1.

When a 3W-1φ circuit originates from a 4W, 3φ, Wye (Y) system only two ungrounded (phase) conductors are used along with a common conductor. The common conductor of a 208/120V 3-wire circuit or system will as stated *"carry approximately the same current as the line-to-neutral load currents of the other conductors"* which is demonstrated per formula and results shown in Figure 310.15(B)(5)(b).1.

COMMON CONDUCTOR COUNTED as "current carrying"

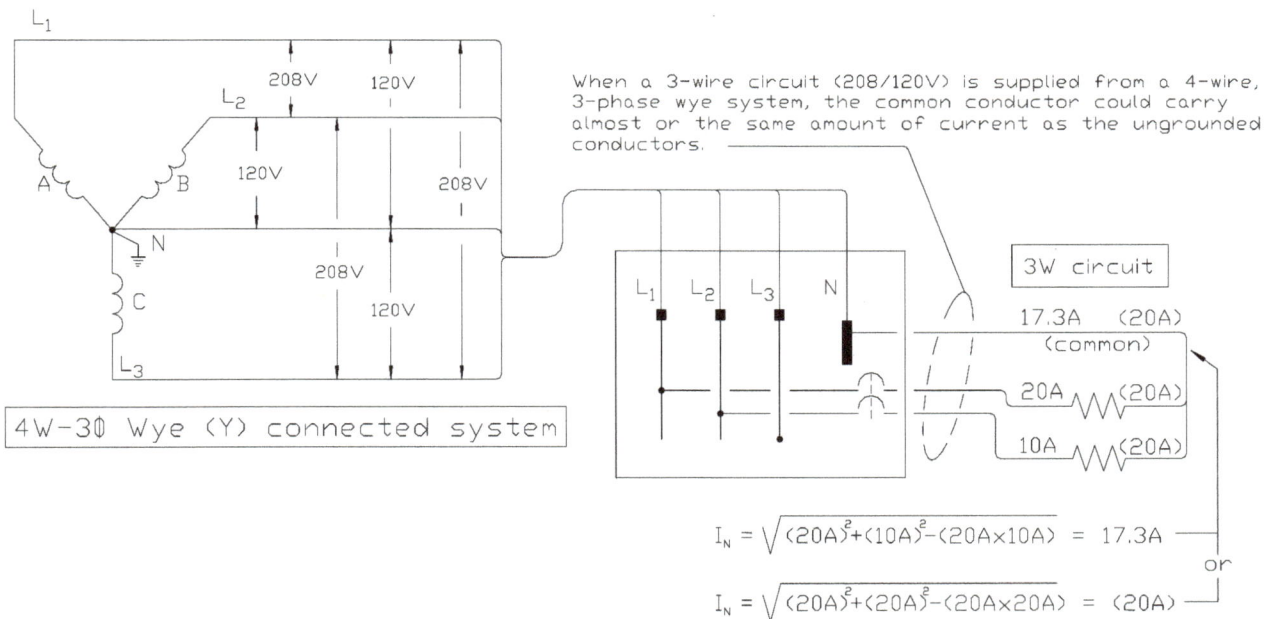

When a 3-wire circuit (208/120V) is supplied from a 4-wire, 3-phase wye system, the common conductor could carry almost or the same amount of current as the ungrounded conductors.

$$I_N = \sqrt{(20A)^2 + (10A)^2 - (20A \times 10A)} = 17.3A$$

or

$$I_N = \sqrt{(20A)^2 + (20A)^2 - (20A \times 20A)} = (20A)$$

310.15(B)(5)(b).1 - 3W circuit supplied from 4W-3φ Wye (Y) connected system

310.15(B)(5)(c)

On a 4-wire, 3-phase wye (Y) circuit where the major portion of the load consists of nonlinear loads, there are harmonic currents present in (flowing through) the *neutral conductor*, thereby requiring the *neutral conductor* to be considered as a current-carrying conductor.

NEC 310.15(B)(5)(c) only applies to 4W, 3φ, Wye (Y) connected circuits. In these type circuits third and other odd-order harmonic currents which occur in the line conductors (caused by nonlinear loads) are in phase and add together in the neutral conductor unlike the Wye systems or circuits in previous illustrations where the loads only consisted of linear loads. See Figures 310.15(B)(5)(c).1 - .4.

harmonic current - Distorted and unwanted nonlinear current that is produced by various types of electronic equipment which consist of electronic devices that operate at frequencies other than 60Hz. These currents can cause excessive amounts of neutral current which can leads to overheating, equipment damage and fires.

NEUTRAL CONDUCTOR (Major Portion [over 50 percent] of load nonlinear) COUNTED as "current-carrying"

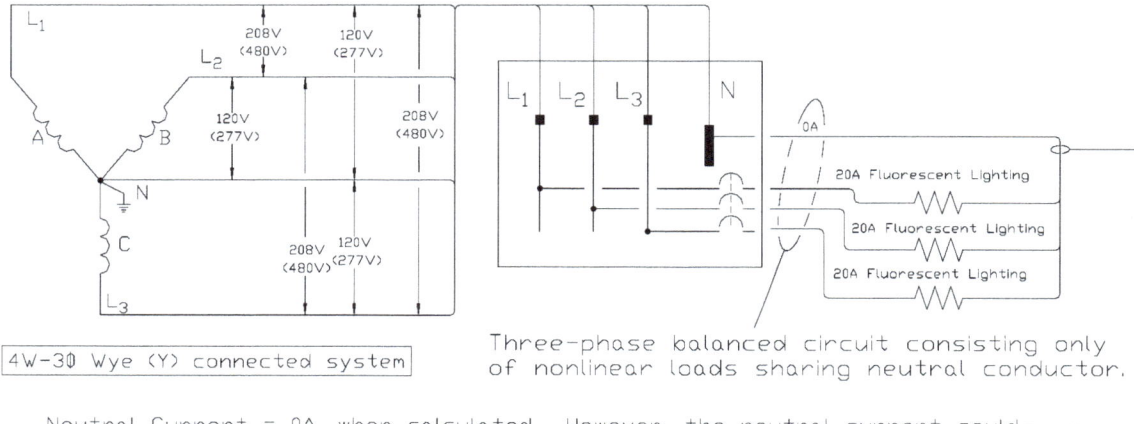

4W-3Ø Wye (Y) connected system

Three-phase balanced circuit consisting only of nonlinear loads sharing neutral conductor.

Neutral Current = 0A, when calculated. However, the neutral current could actually equal up to twice the current flowing through the line conductors. Therefore, where the major portions of loads are nonlinear, the applied formula can not be used to derive the neutral current.

$$I_N = \sqrt{(20A)^2 + (20A)^2 + (20A)^2 - (20A \times 20A) + (20A \times 20A) + (20A \times 20A)} = 0A$$

Figure 310.15(B)(5)(c).1 - 4W, 3φ, Wye (Y) connected system supplying 4W, 3φ, 208Y/120V or 480Y/277V circuits

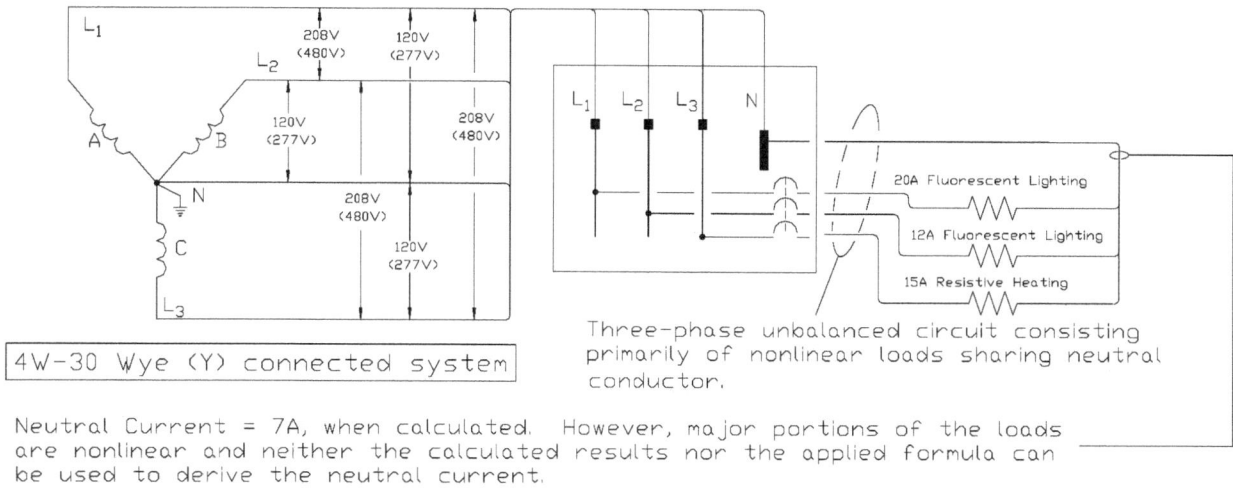

4W-30 Wye (Y) connected system

Three-phase unbalanced circuit consisting primarily of nonlinear loads sharing neutral conductor.

Neutral Current = 7A, when calculated. However, major portions of the loads are nonlinear and neither the calculated results nor the applied formula can be used to derive the neutral current.

$$I_N = \sqrt{(20A)^2 + (12A)^2 + (15A)^2 - (20A \times 12A) + (12A \times 15A) + (20A \times 15A)} = 7A$$

Figure 310.15(B)(5)(c).2 - 4W, 3φ, Wye (Y) connected system supplying 4W, 3φ, 208Y/120V or 480Y/277V circuits

Fundamental Current

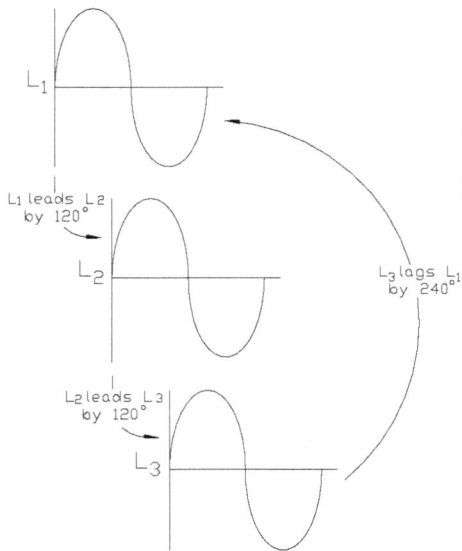

For 4W, 3Ø, 208/120V or 480/277V

When loads consist only of linear loads, harmonic currents are not present, therefore

I_N = "0" amps if loads are balanced

or

the calculated results if loads are unbalanced

Figure 310.15(B)(5)(c).3 - Fundamental current in 4W, 3φ, Wye (Y) connected system

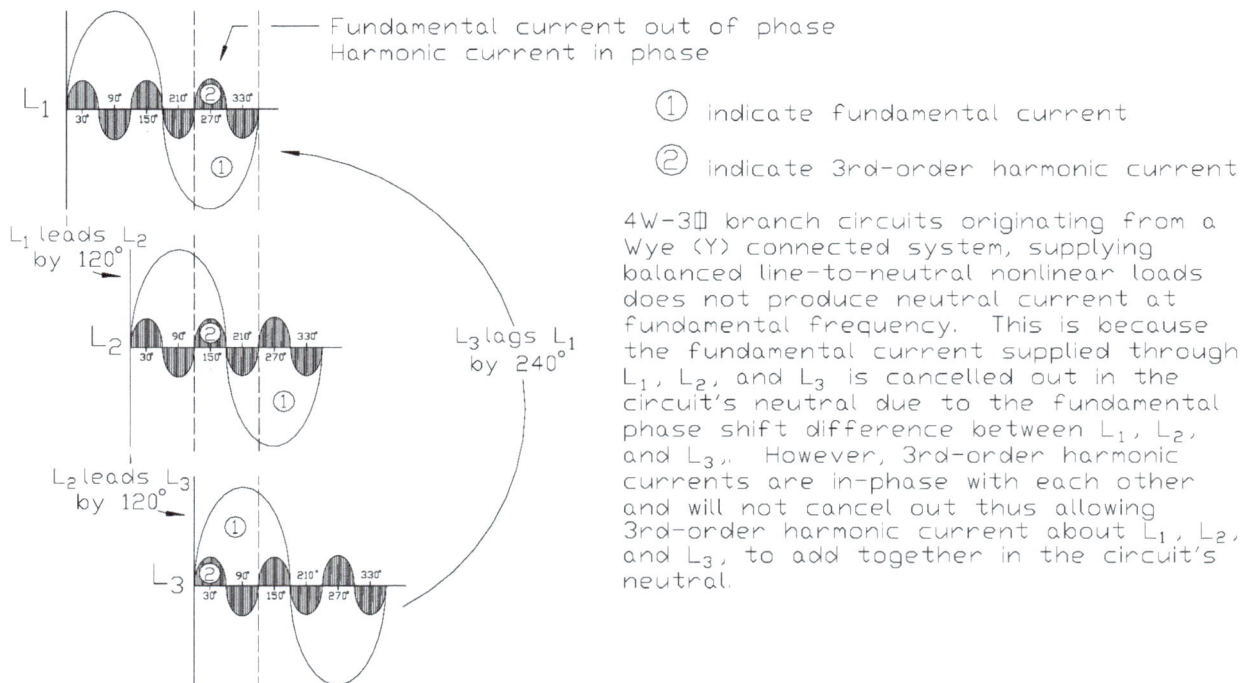

Figure 310.15(B)(5)(c).4 - Fundamental current vs. Harmonic current in 4W,3ϕ, Wye (Y) connected system

For 3W,1ϕ, 240/120V circuits consisting of nonlinear loads whether originating from a 3W, 1ϕ, 240/120V system or from a 4W, 3ϕ, 240/120V Delta (Δ) connected system, harmonic current which occurs in the line conductors are 180° out of phase and cancel each other in the neutral conductor. See Figure 310.15(B)(5)(c).5.

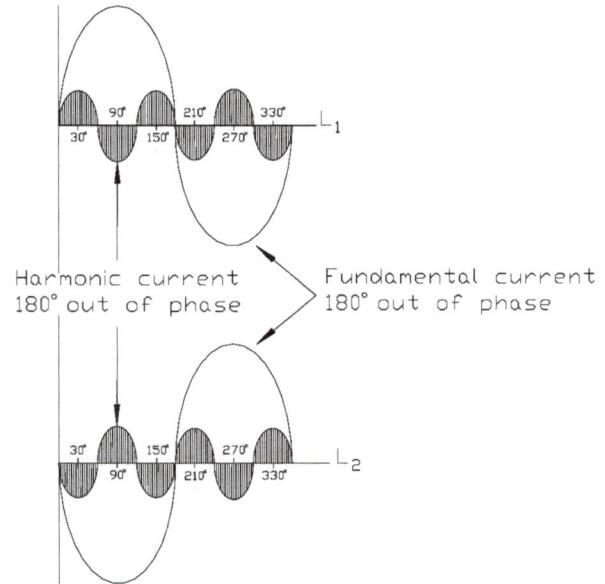

Fundamental and harmonic current flowing
through the line conductors, L_1 and L_2 are
both 180° out of phase which results in
both currents canceling out each other
and eliminating current flow through the
neutral conductor.

Harmonic current
180° out of phase

Fundamental current
180° out of phase

Figure 310.15(B)(4)(c).5 - Fundamental 60Hz current vs. Harmonic current in 3W-1φ system

28. A three phase, 4W, 208/120V, multiwire, lighting branch circuit is installed in intermediate metal conduit to supply a balanced incandescent lighting load in an area where the ambient temperature is 30°C. If 12 AWG THW copper conductors are used as the branch-circuit conductors, determine the maximum allowable ampacity of the conductors.

Unlike an inductive lighting load such as fluorescent, high pressure sodium, mercury vapor, metal halide, etc. where the use of either a ballast, transformer or autotransformer is required to energize a connected lamp or bulb through a light fixture, an incandescent lighting load is one where a light fixture directly energizes a regular light bulb or group of light bulbs. As earlier displayed through illustrated figures an inductive lighting load is classified as a *nonlinear* load while an incandescent lighting load would be classified as a *linear* load.

Because an incandescent lighting load is classified as a linear load the neutral conductor of a multiwire branch circuit (as such) is considered noncurrent-carrying. As a result, only 3 wires (conductors) of the 4 wire circuit are considered current-carrying. Considering the area where these conductors will be installed (30°C) and only 3 current-carrying conductors, the allowable ampacity of the 12 AWG conductors can be taken directly from Table 310.15(B)(16) which is 25 amps.

310.15(B)(6) - Grounding or Bonding Conductor

29. Three 350 kcmil THHN copper conductors are installed in a 3" raceway along with a 6 AWG THHN copper equipment grounding conductor. The conduit is routed through areas where the ambient temperatures are 71°F, 88°F, 108°F, 119°F, and 131°F. Determine the allowable ampacity of the conductors.

The given ambient temperatures (71°F, 88°F, 108°F, 119°F and 131°F) either exceeds or falls below 86°F per Table 310.15(B)(2)(a). Referring to the Correction Factors of Table 310.15(B)(2)(a) per conductor temperature rating - the correction factors at,

$$71°F = \underline{1.04} \quad 88°F = \underline{.96} \quad 108°F = \underline{.87} \quad 119°F = \underline{.82} \quad 131°F = \underline{.76}$$

NEC 310.15(A)(2) states that when two or more calculated ampacities apply to the length of a given circuit, the lower ampacity value shall be used for the entire circuit.

Since the highest ambient temperature reflects the lowest correction factor, and the lowest correction factor per Table 310.15(B)(2)(a) [when multiplied by the ampacity of the conductor] will produce the lowest ampacity, the correction factor to be used in this situation is .76.

The number of conductors enclosed in the 3" raceway is 4, but the question does not imply whether they are current-carrying or not.

However, since one of these conductors is an equipment grounding conductor, NEC 310.15(B)(6) states that equipment grounding conductors are not counted when applying the provisions of 310.15(B)(3)(a). Whether the remaining conductors are current carrying or not the ampacity of these conductors are not required to be adjusted because there are only 3 conductors to be considered. Keep in mind that the equipment grounding conductor should only be taken into consideration when performing raceway fill calculations.

The ampacity of the 350 kcmil THHN copper conductors is 350 amps at 90°C. Using the given ampacity and the .76 correction factor where,

$$350 \text{ amps x } .76 = 266 \text{ amps}$$

The allowable ampacity of the conductors is 266 amps.

310.15(B)(7)/Table 310.15(B)(7) - Conductor Types and Sizes for 120/240Volt, 3-Wire, Single-Phase Dwelling Services and Feeders (30. - 33.) [4]

The conductors listed in Table 310.15(B)(7) where used to supply individual dwelling units of one-family, two-family and multifamily dwellings, shall be permitted as 120/240-volt, 3-wire, single-phase service-entrance (overhead) conductors, service-lateral (underground) conductors, and feeder conductors that serve as the main power feeder to each dwelling unit and are installed in raceway or cable with or without an equipment grounding conductor. For provisions that cover requirements for grounded conductors refer to NEC 215.2, 220.61 and 230.42.

30. Type THWN copper is being used as the service-entrance conductors for a single-family dwelling. The dwelling's electrical service is rated for 110 amperes at 240/120V. Determine the minimum size service-entrance conductors needed for this 3-wire service.

Referring to Table 310.15(B)(7), with a 3-wire, 240/120V electrical service rated for 110 amps, the minimum size service-entrance conductors needed for this service are 3 AWG copper conductors.

31. Determine the minimum size service rating and USE ungrounded service-lateral conductors permitted when a 240/120V three-wire service is being used for a one-family residence. The residence's calculated electrical load is 31.466kVA.

Determine the amperage of the 31.466kVA electrical load at 240 volts.

$$I = \frac{31,466VA}{240V} = 131.11 \text{ amps}$$

At 131.11 amps, according to Table 310.15(B)(7), as a minimum a 150 amps service rating is required with the use of either 1 AWG copper or 2/0 AWG aluminum USE (service lateral) conductors. See Article 338 for the use, installation and construction of this type cable.

32. Determine the maximum size feeder rating permitted when 4/0 AWG copper conductors are used as the service-entrance conductors for a 120/240V, 3-wire single-phase dwelling.

According to Table 310.15(B)(7), the maximum size feeder rating permitted based on the information provided in the question is 250 amps.

33. Determine the minimum size service rating and 75°C ungrounded service conductors needed for a one-family dwelling when a single-phase, three-wire 240/120V electrical service is used. The calculated demand load for this dwelling is 104kVA.

Determine the amperes rating of the 104,000VA demand load at 240 volts.

$$I = \frac{104,000VA}{240V} = 433.33 \text{ amps}$$

Because the data listed in Table 310.15(B)(7) is only applicable for electrical services up to 400 amps, the service rating can be sized based on NEC 230.90(A), *Exception No. 2* and the ungrounded service conductors per Table 310.15(B)(16). Based on the calculated load as a minimum, a 450A service is required.

For this application the rating of the service disconnecting means per NEC 230.79 and the ampacity of the ungrounded service conductors per NEC 230.42(A) must not be less than determined in accordance with Part III, IV, or V of Article 220. Although the demand load (104kVA) for this dwelling was given, if it had to be determined based on a given combination of individual loads, the provisions of Article 220 would have been applied. Therefore, based on the calculated load (433.33A) as a minimum, a 450A service rating is required. Again, based on NEC 230.90(A), *Exception No. 2* which reference the compliance of NEC 240.4(B), the minimum size ungrounded service conductors can be either 600 kcmil copper (420 amps) or 900 kcmil aluminum (425 amps) conductors at 75°C.

Even if multiple service panelboards (load centers) with a combined 450A service rating were used instead for example, two 225A panelboards with main breakers or one panelboard with a 250A main breaker along with a 200A panelboard with a main breaker, etc., the service conductors supplying the service meter would still be required to be either of the sizes mentioned (600 kcmil copper or 900 kcmil aluminum conductors). However, the service conductors being supplied from the service meter to either combination of service panelboards could be sized in accordance with Table 310.15(B)(7) per service panelboard (225A[2]-3/0 AWG copper or 250 kcmil AWG aluminum) *or* (250A-4/0 AWG copper or 300 kcmil AWG aluminum and 200A-2/0 AWG copper or 4/0 AWG aluminum).

FORMULAS FOR ADJUSTING CONDUCTOR AMPACITY

1. When the ampacity of a conductor requires adjusting based on adverse conditions, that is,

    (a) conflicting or favorable ambient temperatures (above or below 30°C/86°F) *and*

    (b) current-carrying conductors enclosed in raceway, cable or directly buried exceeds a total of three,

refer to,

FORMULA 1

    Adjusted Ampacity = **Table Ampacity** x **TCF** (if needed) x **AF** (if needed)

      where,

        Table Ampacity = Ampacity of a conductor as listed in Table 310.15(B)(16) *or* the applicable table per temperature rating, conductor type and size

        TCF = Temperature Correction Factor based on Table 310.15(B)(2)(a) *or* the applicable table

        AF = Adjustment Factor based on Table 310.15(B)(3)(a); the provisions of NEC 310.15(B)(5) must be applied where applicable

2. To determine the required ampacity of a conductor based on the conductor supplying a given load in an adverse condition,

refer to,

FORMULA 2

$$\text{Required Ampacity} = \frac{I^*}{\text{TCF (if needed) x AF (if needed)}}$$

where,

I = Load current

TCF = Temperature Correction Factor based on Table 310.15(B)(2)(a) *or* the applicable table

AF = Adjusted Factor based on Table 310.15(B)(3)(a); provisions of NEC 310.15(B)(5) must be applied where applicable

*Use either given or calculated load, full-load current (FLC) of motor per Article 430 or ampacity of conductor based on Table 310.15(B)(16) *or* applicable table.

3. Refer to NEC 210.19(A)(1), 210.20(A), 215.2(A)(1), 215.3, 240.4(B) and (C). To determine ampacity of conductor and overcurrent device (OD) based on the conductor being used for *continuous* or *noncontinuous* loads or both, first determine the load using **FORMULA 3**, then apply **FORMULA 2**, where applicable.

FORMULA 3

$$\underline{\quad}\text{A (continuous load)} \times 1.25 + \underline{\quad}\text{A (non-continuous loads)} = \frac{\underline{\quad}\text{A}}{\text{(Calculated Load)}^{**}}$$

** Also used to determine size of overcurrent device (OD) based on calculated load and NEC 210.20(A), 215.3, 240.4(B) and (C).

Sample Questions - Adjusting Conductor Ampacity (34. - 40.) [7]

34. Four 2W-120V circuits are installed in a raceway to supply four individual loads. There is a possibility that the ambient temperature where these circuits will be installed could reach 53°C. Determine the required ampacity of each circuit conductor, if the conductors are copper with type ZW insulation and the operating load of each circuit is 13.43A, 18.65A, 23.9A and 27.62A.

The four circuits in this question equates to 8 (4 x 2) current-carrying conductors. Referring to Table 310.15(B)(2)(a), a 70 percent (.70) adjustment factor is required. Considering the insulation type, a .67 correction factor is used per Table 310.15(B)(2)(a). With the given information **FORMULA 2** can be applied to determine the required ampacity of each conductor per load.

$$\text{Required Ampacity} = \frac{13.43 \text{ amps}}{.67 \times .70} = 28.64 \text{ amps}$$

$$= \frac{18.65 \text{ amps}}{.67 \times .70} = 39.77 \text{ amps}$$

$$= \frac{23.9 \text{ amps}}{.67 \times .70} = 50.96 \text{ amps}$$

$$= \frac{27.62 \text{ amps}}{.67 \times .70} = 58.89 \text{ amps}$$

If sizing conductors per required ampacity, the following conductors (Type ZW) would be required, 10 AWG (28.64A), 8 AWG (39.77A) and 6 AWG (50.96A) and (58.89A).

35. Four THW aluminum conductors are installed in a 1½" conduit located in an area where the ambient temperature is 98°F. These conductors supply a 77 amps load and are all current-carrying. Determine the minimum size conductors and overcurrent device required for this installation.

Based on an ambient temperature of 98°F, the correction factor for the aluminum conductors is .88. A .80 adjustment factor is used based on 4 current-carrying conductors. With a load current of 77 amps, **FORMULA 2** is used to determine the size conductor required for this installation.

$$\text{Required Ampacity} = \frac{77 \text{ amps}}{.88 \times .80} = 109.38 \text{ amps}$$

Based on the calculated results, a 1/0 AWG THW aluminum conductor which has an ampacity of 120 amps at 75°C must be used.

As a minimum, a device with an 80A rating can be used. Remember the overcurrent device can be sized based on the given 77 amps load and not the required ampacity of the conductor, 109.38 amps. Refer to NEC 210.20(A) and 215.3. However, based upon the ampacity of the aluminum conductor, an overcurrent device rated for 125A can be used per NEC 240.4(B).

36. Eight THWN copper conductors are used to supply separate 277V mercury vapor outdoor lighting circuits that are energized for only 2.5 hours at most. Each line conductor pulls a total of 24.7 amps. The conductors originate in an outdoor panelboard and then are ran equally in two separate runs of PVC that extends 4 feet beneath the panelboard before being ran underground. The surrounding temperature has been known to reach up to 95°F. What size conductors and overcurrent device are required for these circuits?

Based on an ambient temperature of 95°F, the correction factor for THWN copper conductors is .94. Although there are eight conductors mentioned in the question, only 4 conductors per raceway need be considered for derating. The question states that the conductors are distributed equally in separate raceways. Because the type loads these conductors will supply are *nonlinear*, the neutral conductor is considered along with the line conductors as being current-carrying. Therefore, the number of current-carrying conductors is 4 and requires adjusting.

However, according to NEC 310.15(B)(3)(a)(3), because the conductors are protected by means of the specified raceway [Rigid polyvinyl chloride conduit (PVC)] and the raceways do not exceed 10 feet (above ground) and contains no more than 4 conductors the conductors are not required to be derated per Table NEC 310.15(B)(3)(a).

With a load current of 24.7 amps, **FORMULA 2** is used to determine the size conductors required for this installation.

$$\text{Required Ampacity} = \frac{24.7 \text{ amps}}{.94} = 26.28 \text{ amps}$$

Based on the calculated results, as a minimum a 10 AWG THWN copper conductor which has an ampacity of 35 amps at 75°C must be used.

NEC 240.4(D)(7) states that the overcurrent device for a 10 AWG copper conductor shall not exceed 30 amps. Therefore, in this situation either a 25 or 30 amps overcurrent device can be used however, the 25 amps overcurrent device may cause nuisance tripping. Remember the overcurrent device can be sized based on the given load at 24.7 amps and not the required ampacity of the conductor, 26.28 amps.

37. Consider question No. 35. if the circuits are operated continuously. What size conductors and overcurrent device are now required for this circuit?

Because the conductors will be used continuously, **FORMULA 3** can be applied and then **FORMULA 2**. Using the given information in question No. 35.,

$$77A \times 1.25 = 96.25A$$

Considering the results of the calculation (96.25A), a 100 amps overcurrent device is required. Using the calculated results, the required ampacity of the circuit conductors can be determined using **FORMULA 2**,

$$\text{Required Ampacity} = \frac{96.25 \text{ amps}}{.88 \times .80} = 136.72 \text{ amps}$$

At 136.72 amps, as a minimum, 3/0 AWG THW aluminum conductors rated for 155 amps are required.

38. Consider question No. 36., if the circuits operated continuously. What size conductors and overcurrent device are now required for these circuits?

Because the conductors will be used continuously, **FORMULA 3** can be applied and then **FORMULA 2**.

Using the given information in question No. 36.,

$$24.7A \times 1.25 = 30.88A$$

Considering the results of the calculation (30.88A), a 35 amps overcurrent device is required. Using the calculated results, the required ampacity of the circuit conductors can be determined using **FORMULA 2**,

$$\text{Required Ampacity} = \frac{30.88 \text{ amps}}{.94} = 32.85 \text{ amps}$$

Based on the calculated results, a 10 AWG THWN copper conductor may appear as if it can be used for this installation also. However, NEC 240.4(D)(7) states that the maximum size overcurrent device that can be used with a 10 AWG conductor is a 30A device. Therefore, an 8 AWG THWN copper conductor must be used in this situation based on the use of a 35A overcurrent device.

39. Two 480V, 3-phase motors rated for 25 and 40 horsepower are being installed on the rooftop of a commercial building to replace two similar motors. During previous summer months the replaced motors were exposed to temperatures that peaked at 127°F. Because the branch-circuit conductors supplying these motors were undersized the motor's life-time expectancy was shorten. New motor branch-circuit conductors will be installed in the same 1½" EMT raceway that served the replaced motors. If the raceway is mounted 15" above the building's rooftop and type THWN-2 copper conductors were previously used, what size conductors are now required?

According to NEC 430.6(A)(1), the ampacity of motor conductors must be determined based on the full-load current values of the applicable tables of that article. Also NEC 430.22 requires all branch-circuit conductors supplying a single motor with a continuous duty application to have an ampacity not less than 125 percent of the motor's full-load current. Based on these requirements the rated ampacity of each motor's branch-circuit conductor's as a minimum must be,

25HP, 480V, 3-phase - 34A x 1.25 = 42.5A
40HP, 480V, 3-phase - 52A x 1.25 = 65A

Considering the correction factors of Table 310.15(B)(2)(a) and a type THWN-2 copper conductor being exposed to an ambient temperature of 127°F, the ampacity of the conductor must be reduced to seventy-six percent (.76) of its rated ampacity. However, per NEC and Table 310.15(B)(3)(c) when conduits are exposed to sunlight on rooftops the adjustments shown in Table 310.15(B)(3)(c) must be added to the outdoor temperature to determine the applicable ambient temperature for application of the correction factors in Table 310.15(B)(2)(a).

Referring to the information given in the last row of Table 310.15(B)(3)(c), when a conduit is mounted 15" above the rooftop to the bottom of the conduit a temperature adder of 25°F is added to the given ambient temperature. As a result, the combined temperatures amount to 152°F (127°F + 25°F) which renders a correction factor of fifty-eight percent (.58) per Table 310.15(B)(2)(a).

Applying the provisions of Table 310.15(B)(3)(a), where six current-carrying motor conductors are being installed in the same 1½" metal raceway the ampacity of the conductors must be reduced to eighty percent (.80).

Based on the given information, since the motor conductors will be used continuously, **FORMULA 3** can be applied followed by the application of **FORMULA 2** to determine the size motor conductors needed to supply each motor load.

Per **FORMULAS 2 and 3**

$$25HP, 480V, 3\text{-phase} - \frac{42.5A}{.58 \times .80} = 91.6A$$

$$40HP, 480V, 3\text{-phase} - \frac{65A}{.58 \times .80} = 140.1A$$

Based on the above calculations, the 25HP motor will require 4 AWG THWN-2 (95A) copper conductors and the 40HP motor will require 1 AWG THWN-2 (145A) copper conductors per Table 310.15(B)(16).

40. A 4W, 3φ, 208/120V feeder installed in metal raceway will be used to feed a 3-phase panelboard. The panelboard will supply five continuous loads of 32 amps, 45 amps, 59 amps, 65 amps and 81 amps along with a combined non-continuous load of 186.22 amps. All loads are considered nonlinear. The feeder conductors will be routed through an extended area where the temperature at worst reaches 44°C. Determine the required ampacity of the feeder conductors if both copper and aluminum RHW conductors are considered. What size overcurrent device is required?

In this situation a four-wire feeder consisting of three line conductors and one grounded neutral conductor is considered. Because the feeder will be supplying a panelboard that will supply nonlinear loads the grounded neutral conductor is considered a current carrying conductor per NEC 310.15(B)(5)(c). Applying **FORMULA 3** the continuous and non-continuous loads totals 538.72A ([32A + 45A + 59A + 65A + 81A] x 1.25 + 186.22A).

Being that the four (4W) feeder conductors are installed in metal raceway and all conductors are current-carrying the rated ampacity of the conductors is required to be reduced by eighty percent (.80) per Table 310.15(B)(3)(a).

Because the feeder conductors will be routed through an extended area where the temperature at worst reaches 44°C, the required correction factor per Table 310.15(B)(2)(a) is eighty two percent (.82) for both copper and aluminum RHW conductors.

As a minimum, the required ampacity of the feeder conductors based on the application of **FORMULA 2** amounts to,

$$\frac{538.72A}{.80 \times .82} = 821.22A$$

Because there are no single conductors listed in Table 310.15(B)(16) which has a rated ampacity of 821.22A, two sets of parallel conductors will be used where each conductor must have as a minimum a rated ampacity of 410.61A (821.22A/2). Based on 410.61A, as a minimum 600 kcmil RHW copper conductors must be used which are rated for 420A or 900 kcmil RHW aluminum conductors must be used which are rated for 425A.

In sizing the feeder's overcurrent protection, NEC 215.3 requires the overcurrent device to be rated no less than the noncontinuous load plus 125 percent of the continuous load. Based on previous calculations, a 600A overcurrent device is required based on 538.72A.

Also see question No. 7., Article 210.

FORMULAS FOR CALCULATING NEUTRAL CURRENT

In most cases the neutral conductors of an electrical systems carries only the unbalanced portion of a load. When 3W-1ϕ or multiwire (4W-3ϕ) circuits (whether balanced or unbalanced) originates from a 4W-3ϕ Wye (Y) connected system or source consisting of nonlinear loads, the neutral conductor under most situations could possibly encounter up to 200 percent of the current flowing through the phase conductor with the largest load. Because of this the amount of current that could flow through the neutral conductor when *nonlinear* loads are present cannot be calculated.

On the other hand, when a 3W-1ϕ or a multi-wire circuit originates from a 4W-3ϕ Wye (Y) connected system or where a 4W-3ϕ Wye (Y) connected source consist exclusively of *linear* loads the following formula is used to provide an approximate means for calculating the neutral current,

$$I_N = \sqrt{(I_{L1})^2 + (I_{L2})^2 + (I_{L3})^2 - (I_{L1} \times I_{L2}) + (I_{L2} \times I_{L3}) + (I_{L3} \times I_{L1})}$$

where,

I_N = current in the Neutral
I_{L1} = current in Line 1 I_{L2} = current in Line 2 I_{L3} = current in Line 3

multiwire branch circuit - A branch circuit consisting of two or more ungrounded conductors having a potential difference between them, and a grounded conductor having equal potential difference between it and each ungrounded conductor of the circuit and that is connected to the neutral or grounded conductor of the system.

For 3W, 1ϕ, 240/120V circuits such as those originating from 3W, 1ϕ, 240/120V systems or from 4W, 3ϕ, 240/120V Delta (Δ) connected systems the following formula is provided for calculating neutral current,

$$I_N = I_{L1} - I_{L2}$$

where,

$I_{L1 \text{ and } L2}$ = the applicable line currents

Examples Referencing Neutral Current (41. - 46.) [6]

41. A 4W-3ϕ, 208/120V circuit supplies the following neutral loads:

> L_1 to N = (18A) Incandescent lighting (linear)
> L_2 to N = (26A) Motors (linear)
> L_3 to N = (13A) Fluorescent lighting (non-linear)

How much current will flow through the neutral? Is the neutral conductor considered current carrying?

Although the major portion of the above loads is linear, there is no simple formula for calculating the neutral current for circuits consisting of linear and nonlinear loads. In order to use the given formula for multiwire circuits originating from 4W-3ϕ Wye (Y) connected systems, circuits must only consist of linear loads. As a result the neutral current cannot be calculated by such formula.

Because the major portion of the above loads are linear, according to the provisions of NEC 310.15(B)(5)(c) the neutral conductor is not considered current-carrying.

42. A 3W-1ϕ, 208/120V circuit is used to supply an unbalanced load where 18A flows through one line conductor and 12.3A through the other. How much current will the neutral (common) conductor carry? Is the conductor considered current-carrying?

$$L_1 = 18A \quad L_2 = 12.3A$$

$$I_N = \sqrt{(18A)^2 + (12.3A)^2 - (18A \times 12.3A)}$$

$$= \sqrt{324A^2 + 151.29A^2 - 221.4A^2}$$

$$= \sqrt{253.89A}$$

$$= 15.93A$$

Based on the above calculation, 15.93A will flow through the neutral conductor. Although it might not seem possible for the neutral to carry this amount of current when compared to that of line (ungrounded) conductors, the formula and results of the calculation substantiates that stated in NEC 310.15(B)(5)(b) in reference to a common conductor. As a result, the neutral (common) conductor must be counted as a current-carrying conductor.

43. How much current flows through the neutral conductor of a balanced 67 amps 4W-3φ 480/277V high-pressure sodium lighting load? Is the neutral conductor considered current carrying?

Because high-pressure sodium lights are nonlinear and capable of producing harmonic currents the amount of current flowing through the neutral conductor could reach up to 200 percent of the line current, that is, 134A.

According to NEC 310.15(B)(5)(c), the neutral conductor is considered current-carrying when the major portion of the load consist of nonlinear loads.

44. A 240/120V-1φ multiwire branch circuit supplies a dishwasher and garbage disposal. The dishwasher pulls 9.6 amps and the garbage disposal pulls 3.8 amps. How much current will the neutral conductor carry?

When a 240/120V-1φ multiwire branch circuit originates from a 240/120V-1φ system the neutral conductor will only carry the unbalanced current between the two line conductors, that is,

$$I_N = 9.6A - 3.8A = 5.8A$$

45. Calculate the neutral current for a 4-wire, 208Y/120V feeder where all line loads equal 55A.

When calculated,

$$I_N = \sqrt{(55A)^2 + (55A)^2 + (55A)^2 - (55A \times 55A) + (55A \times 55A) + (55A \times 55A)}$$

$$= 0 \text{ amps}$$

The neutral current in this situation amounts to zero, based on all line loads being linear.

46. Calculate the neutral current for a 4-wire, 208Y/120V feeder where $L_1 = 33.8A$, $L_2 = 46.3A$ and $L_3 = 38.7A$.

When calculated,

$$I_N = \sqrt{(33.8A)^2+(46.3A)^2+(38.7A)^2 - (33.8A \times 46.3A)+(46.3A \times 38.7A)+(33.8A \times 38.7A)}$$

$$= \sqrt{4783.82A - 4664.81A}$$

$$= \sqrt{119.01A}$$

$$= 10.91A$$

The neutral current in this situation amounts to 10.91 amps, where again all line loads are considered linear.

ARTICLE 312 - Cabinets, Cutout Boxes, and Meter Socket Enclosures

WIRE BENDING SPACE IN CABINETS, CUTOUT BOXES and METER SOCKET ENCLOSURES

Where conductors enter or leave cabinets, cutout boxes and meter socket enclosures the possibility of damaging the conductor's insulation is always present. For larger conductors the possibility is certainly true because of the lack of flexibility, weight and awkwardness that's involved with handling larger conductors. NEC and Tables 312.6(A) and (B) provide specifications which outlines gutter width, wire-bending space and how conductors are required to be bent (deflected) in mentioned enclosures. The purpose of these provisions is to prevent damaging the conductor or the conductor's insulation during installation.

For best results, cabinets, cutout boxes and meter socket enclosures should be installed such that the bending of conductors is minimized where they exit from the conduit. Bends should always be sweeping and finessed rather than sharp and abrupt.

cabinet - A surface or flush mounted enclosure that's equipped with either a vertical or horizontal hinged swinging door or a surface or flush mounted cover where circuit breakers are likely used. See Figure 312.

cutout box - A surface-mounted enclosure that's equipped with either a vertical or horizontal hinged swinging door where fuses are likely used.

meter socket enclosure (can) - An enclosure that's used to accommodate electric watt-hour meters and service conductors.

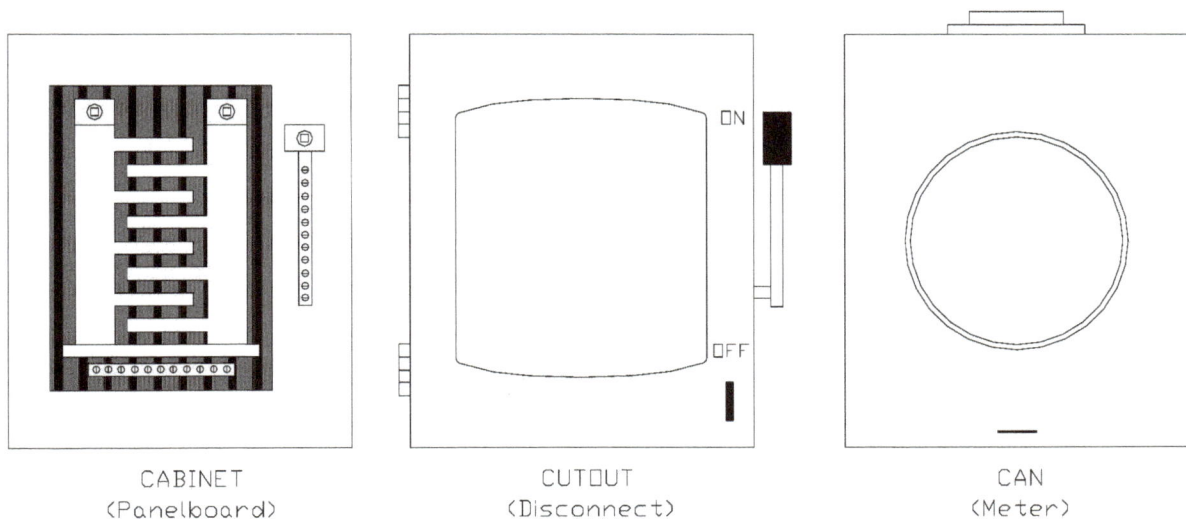

CABINET (Panelboard)   CUTOUT (Disconnect)   CAN (Meter)

Figure 312 - Cabinet, Cutout boxes and Meter Socket Enclosures

Cabinet, cutout box or meter can gutter space refers to the open space within an enclosure that does not include terminals, overcurrent devices or switches. Gutter space in cabinets, cutout boxes or meter cans depending on the type conductor bend is referenced in NEC 312.6(A) and (B).

NEC 312.6(A) covers the minimum gutter width needed on all sides of an enclosure for bending (deflecting) conductors per Table 312.6(A).

NEC 312.6(B) covers the minimum wire bending space needed for making connections to circuit breakers, fuseholders or busbar terminals.

NEC 312.6(B)(1) states that Table 312.6(A) must be used for selecting the minimum wire bending space when conductors do not enter or leave an enclosure through the wall opposite its terminals. Where conductors enters or leaves enclosures through the wall opposite its terminal, NEC 312.6(B)(2) reference Table 312.6(B) for selecting minimum wire bending space. The minimum wire bending space in both tables is determined based on the number of conductors being used per terminal.

WIRE BENDING SPACE PROCEDURES

When a conductor or conductors do not enter or leave an enclosure through the wall opposite its lugs or terminals, refer to Table 312.6(A) for minimum wire bending space and gutter width. Observe Figure 312.6(A).

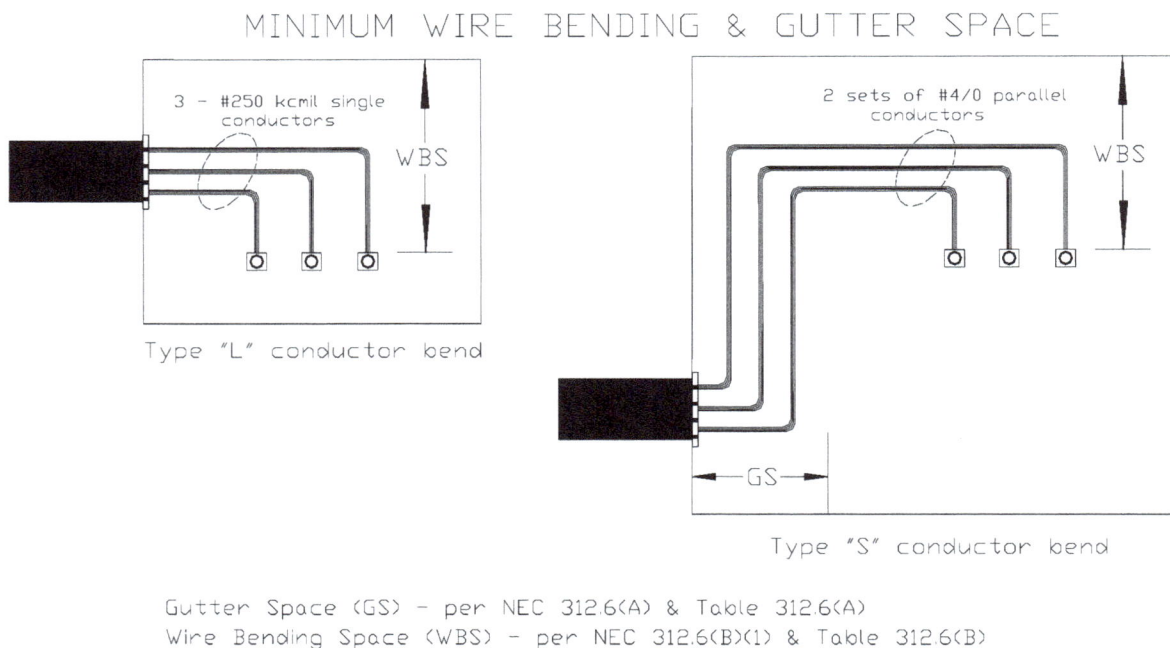

MINIMUM WIRE BENDING & GUTTER SPACE

Gutter Space (GS) – per NEC 312.6(A) & Table 312.6(A)
Wire Bending Space (WBS) – per NEC 312.6(B)(1) & Table 312.6(B)

Figure 312.6(A) - Conductor entrance not opposite terminal

When a conductor or conductors enters or leaves an enclosure through the wall opposite its lugs or terminals refer to Table 312.6(B) for minimum wire bending space. Observe Figure 312.6(B).

MINIMUM WIRE BENDING & GUTTER SPACE

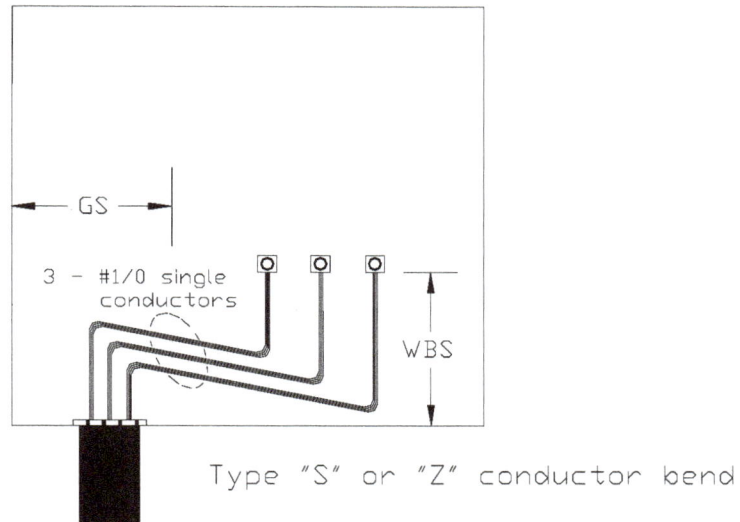

Gutter Space - per NEC 312.6(A) & Table 312.6(A)
Wire Bending Space (WBS) - per NEC 312.6(B)(2) & Table 312.6(B)

Figure 312.6(B) - Conductor entrance opposite terminal

312.6(B)(1) - Conductors Not Entering or Leaving Opposite Wall (Wire-Bending Space at Terminals) (1. - 3.) [3]

1. Refer to Figure 312.6(A). Determine the minimum wire bending space needed to accommodate the Type "L" conductor bend.

In accordance with NEC 312.6(B)(1), where conductors do not enter or leave an enclosure through the wall opposite its terminals, the wire bending space of such an enclosure is determined per Table 312.6(A).

Because only single conductors terminate on each individual terminal, Column 1 (W/T) of Table 312.6(A) is referenced.

For this particular installation the minimum wire bending space must be 4½" when 250 kcmil single conductors are installed per terminal.

2. Refer to Figure 312.6(A). Determine the minimum wire bending space and gutter width needed to accommodate the Type "S" conductor bend.

Again, in accordance with NEC 312.6(B)(1), where conductors do not enter or leave an enclosure through the wall opposite its terminals, the wire bending space of such an enclosure is determined per Table 312.6(A). Table 312.6(A) is also used to determine the gutter width of the enclosure per NEC 312.6(A).

Unlike the previous installation the conductors here are installed in two sets of parallel conductors per individual terminal. Therefore, Column 2 (W/T) of Table 312.6(A) is referenced.

For this particular installation the minimum wire bending space and gutter width must be 6" when two sets of 4/0 conductors are installed in parallel per terminal.

3. Refer to Figure 312.6(B). Determine the minimum wire bending space and gutter width needed to accommodate this installation.

In accordance with NEC 312.6(B)(2), where conductors enters or leaves an enclosure through the wall opposite its terminals the wire bending space of such an enclosure is determined per Table 312.6(B). The gutter width of this enclosure is determined per NEC 312.6(A) and Table 312.6(A).

Because only single conductors terminate on each individual terminal, Column 1 (W/T) of Table 312.6(A) is referenced to determine the gutter width and Column 1 (W/T) of Table 312.6(B) is referenced to determine the wire bending space.

The minimum gutter width must be 3½" and the minimum wire bending space must be 5½".

ARTICLE 314 - Outlet, Device, Pull, and Junction Boxes; Conduit Bodies; Fittings; and Handhole Enclosures

THE MAXIMUM NUMBER OF CONDUCTORS PERMITTED IN A BOX

Article 314 of the NEC provides guidelines for sizing various types of boxes based on need and application. When a box, designed to accommodate a certain number of conductors is utilized beyond its capacity, safety and the risk of hazards are always compromised. Overcrowding or jamming to many conductors into a box can either nick the conductor's insulation or cause too much stress on the conductor. These conditions can also lead to poor ventilation which causes heat given off by the conductors to buildup and overheat the insulation. Forcing a switch or receptacle into a box crowded with conductors only helps to influence the possibility of a short-circuit or ground-fault occurring.

Table 314.16(A) list the maximum number of conductors permitted in standard size metal boxes. However, the table is only applicable when conductors of the same size are installed in the same box. Table 314.16(B) list the volume of conductors ranging in sizes 18 AWG to 6 AWG.

When conductors of different sizes are installed in the same box, the size of the box must be determined based on the listing of the conductor's volume referenced in Table 314.16(B).

NEC 314.16 and affiliated sections defines in detail the methods for calculating the volume of boxes or conduit bodies that are used to enclose conductors of sizes 6 AWG and smaller.

Because all dimensional guidelines discussed in the NEC are minimal, it is often good practice to use boxes that are larger than the minimum size. This further helps to eliminate insulation damage and overcrowding, and for future use will provide spare capacity for the installation of additional conductors or devices. Overall, the "larger-than-minimum concept" can be cost effective and reduce labor and material expense.

ACCORDING TO THE NEC

BOXES AND CONDUIT BODIES

The following NEC sections and tables are applicable for understanding how boxes and conduit bodies are sized and selected per Article 314.

NEC 314.16 - Number of Conductors in Outlet, Device, and Junction Boxes, and Conduit Bodies - When selecting outlet boxes, device boxes, junction boxes or conduit bodies for use in any type electrical installation, the maximum number of conductors permitted in each enclosure must first be determined to ensure there is enough space for all enclosed conductors. All boxes and conduit bodies must meet the minimum volume (cubic inches) as provided or calculated. Terminal housings supplied with motors are excluded from this section. Conductors larger than 6 AWG that are required to be enclosed in either a box or conduit body must adhere to NEC 314.28.

NEC 314.16(A) - Box Volume Calculations - The total volume of a box shall include the actual volume of a box combined with the marked volume (in cubic inches) of assembled components such as plaster rings, domed covers, extension rings, and others [See Figure 314.16(A)]. The volume of an extension ring is not required to be marked because its volume is listed in Table 314.16(A). Remember the only difference between a metal box and an extension rings is, a box has one opening and an extension ring has two.

(1) **Standard Boxes** - Table 314.16(A) provides the volumes of standard boxes that are not marked with their volume.

(2) **Other Boxes** - Boxes not listed in Table 314.16(A), and nonmetallic boxes shall be marked by the manufacturer with their volume.

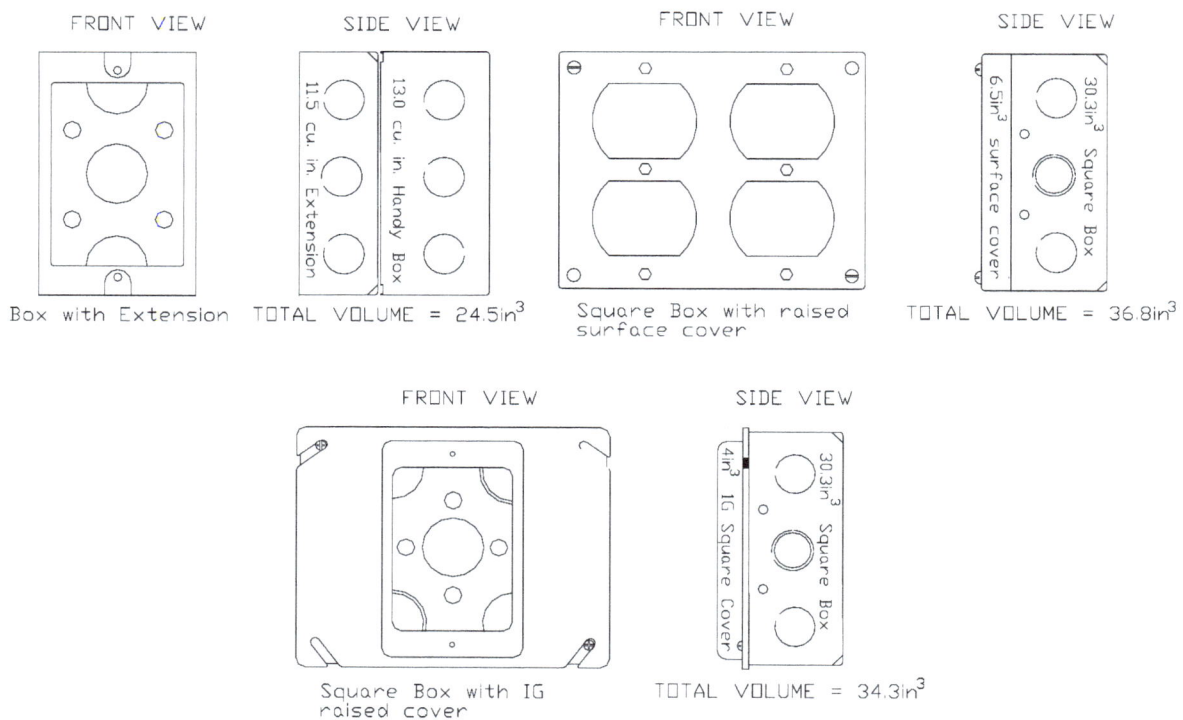

Figure 314.16(A) - Boxes and assembled sections

Table 314.16(A) - Metal Boxes - Table 314.16(A) provides the maximum number of same size conductors, ranging from sizes 18 AWG to 6 AWG, that are permitted in standard metal boxes. There are no allowances in this table for switches, receptacles, fixture studs, fixture hickeys, nor cable clamps. Therefore, if using this table the allowable volume for each mentioned device or component must be deducted as outlined in NEC 314.16(B).

volume - The total cubic inch capacity of an individual box or a combination of the box and an added cover (such as raised covers and extension rings) that will safely house conductors, devices and other permitted components. See Figure 314.16(A).

NEC 314.16(B) - Box Fill Calculations - When a wiring enclosure is needed to enclose conductors up to 6 AWG and the enclosure will contain components such as switches, receptacles, fixture studs, fixture hickeys, and cable clamps, the provisions as outlined in this section must be adhered to. Other type components such as raceway fittings (compression and set-screw connectors), cable connectors, locknuts, bushings, chase nipples, wirenuts and other type splicing or grounding devices are not considered when calculating box fill [See Figure 314.16(B)].

Table 314.16(B) - Volume Required per Conductor - Table 314.16(B) list the volume (cubic inch capacity) of conductors ranging from sizes 18 AWG to 6 AWG when conductors of different sizes are to be installed in the same wiring enclosure. When a wiring enclosure is needed to enclose conductors of different sizes the maximum number of conductors permitted in an enclosure is calculated based on this table.

(C) Conduit Bodies

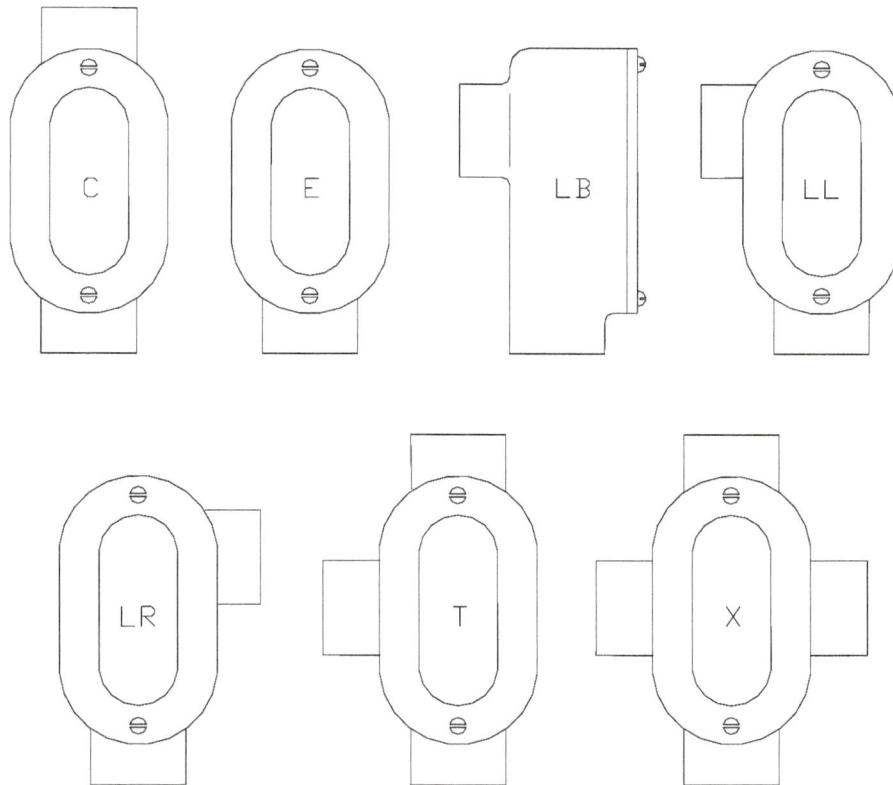

Figure 314.16(C) - Conduit body types

NEC 314.16(C)(1) - General - If conduit bodies, other than short radius types are used to enclose conductors that are 6 AWG and smaller, the cross-sectional area of the <u>conduit body</u> shall not be less than twice the cross-sectional area of the largest conduit or tubing entering the conduit body. The maximum number of conductors permitted in a conduit body shall be determined using the same method as that used for the conduit to which the conduit body will be attached per **Table 1** of Chapter 9.

NEC 314.16(C)(2) - With Splices, Taps, or Devices - Conduit bodies that are marked with their volume are permitted to contain splices and taps. The procedure outlined in NEC 314.16(B) and the conductor volumes listed in Table 314.16(B) are used to calculate the maximum number of conductors permitted in a conduit body. The maximum number of conductors permitted must not exceed the conduit bodies marked volume.

conduit body - A specialized form of an outlet box that has either one or more threaded openings or hubs based on conduit size. These apparatus are made available in a variety of different shapes, types, and openings to satisfy the needs of all electrical installations. Trade names such as condulet, electrolet, unilet or simply conduit fitting are other names used to reference this apparatus. Figure 314.16(C) display different types of conduit bodies most frequently used.

spliced conductor - A conductor that has been joined together with one or more conductors of equal size and/or ampacity by both electrical and mechanical means to provide electrical continuity. See Figure 240.2.

tap conductor - A connection between a smaller conductor and a larger conductor where the smaller conductor is not protected by an overcurrent device according to its ampacity. See Figure 240.2.

NEC 314.16(C)(3) - Short Radius Conduit Bodies - Short radius conduit bodies [see Figure 314.16(C)(3)] enclosing conductors of sizes 6 AWG and smaller shall not contain splices, taps, or devices and shall be of an adequate size to provide free space for all enclosed conductors to eliminate conductor overcrowding and jamming.

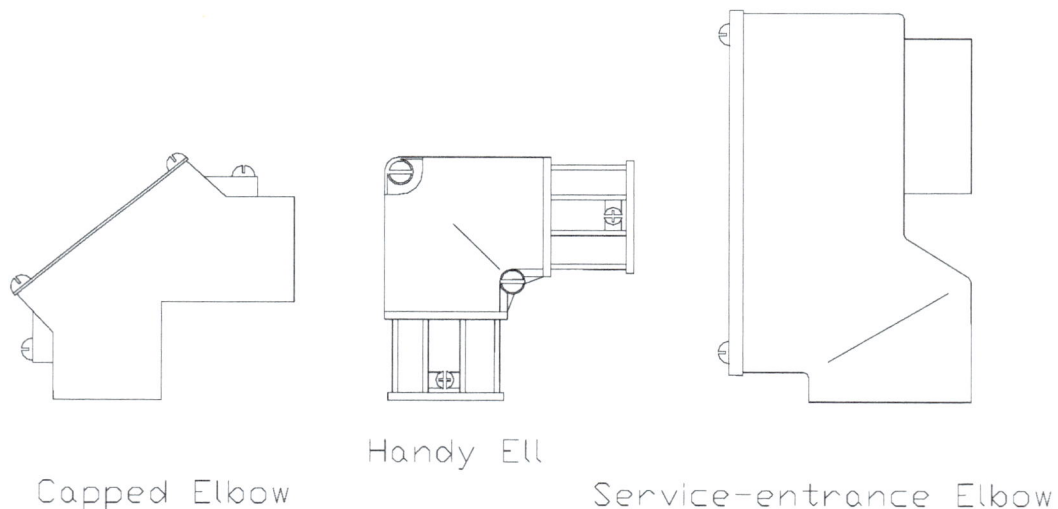

Capped Elbow   Handy Ell   Service-entrance Elbow

Figure 314.16(C)(3) - Short radius conduit bodies

APPLYING THE NEC

BOX FILL CALCULATIONS

When Box Contains Conductors of Same Size

When a metal wiring enclosure such as an outlet box, devce box, junction box, etc., is needed to enclose conductors of the same size regardless of conductor's insulation type, refer to the following procedures,

1. Determine the number of conductors permitted in a selected metal box per Table 314.16(A), Columns 1 (Box Trade Size) and 3 (Maximum Number of Conductors)

or

2. Select the size metal box according to the given number of conductors

Table 314.16(A) - Metal Boxes (1. - 8.) [8]

1. What is the approximate volume of a 4-11/16" x 2-1/8" square box?

By calculation, [4-11/16"(4.6875) x 4-11/16"(4.6875)] x 2-1/8"(2.125) = 46.69 in^3

The approximate volume of a 4-11/16" x 2-1/8" square box is 46.69 in^3.

2. What is the approximate volume of a 4" x 1½ " round box?

By using the formula for a circle,

$$\mathbf{A}_{(area\ of\ circle)} = \boldsymbol{\pi r^2} \text{ (pie [}\pi\text{] x radius squared [}r^2\text{])}$$
where π = 3.14 (rounded off) and r = radius of a circle

the initial step of determining the volume of a round box can be calculated. The 4" measurement of the box identifies the diameter of the box. With this, the radius of the box can be determined by dividing the diameter (d) by 2, that is r = d/2. The radius of the box is, 2".

$$4"/2 = 2" \text{ (inches - in.)}$$

Knowing the radius, the area of the round box can now be calculated.

$$A = 3.14 \text{ x } (2in)^2 = 12.56 \text{ in}^2$$

Because volume (**V**) is a cubic measurement [where in this case cubic inches (in^3)] a third dimension has to be included to determine the volume of the box. This dimension is the 1½" measurement which identifies the depth of the box. In calculating the volume of a circular object the following formula is used,

$$\mathbf{V} = A \text{ x } d \text{ (depth)}$$

As a result,

$$\mathbf{V} = 12.56 \text{ in}^2 \text{ x } 1\frac{1}{2} \text{ (1.5) in} = 18.84 \text{ in}^3.$$

The approximate volume of a 4" x 1½" round box is 18.84 in³.

3. If the minimum (allowable) volume of the 4-11/16" x 2-1/8" square box is 42 in³, determine the allowable percent fill of the box.

Considering the approximate volume of the box, 46.69 in³ and the given minimum volume, 42 in³, approximately 90 percent of the box's approximate volume can be actually utilized, that is,

$$42 \text{ in}^3 \text{ / } 46.69 \text{ in}^3 = .90 \text{ (rounded up)}$$

To determine the allowable percent fill for each metal box listed in Table 314.16(A), divide the box's minimum volume by the approximate volume of the box in question.

4. What is the minimum size square box required to enclose five 10 AWG THW and four 10 AWG TW copper conductors?

There are several boxes that will satisfy this requirement, however to meet the minimum requirement, a 4-11/16" x 1¼" square box (25.5 in³) is selected. This size box has the capacity to hold up to ten 10 AWG conductors. A smaller square box could be used with an extension or plaster ring, etc., depending upon the application, to either meet or exceed the 25.5 in³ requirement. To exceed the minimum requirement without the use of a supplemental part those square boxes surpassing 25.5 in³ must be used.

Unlike raceway fill calculations, only the size of a conductor is taken into consideration when performing box fill calculations. The type insulation that surrounds the core conductor is irrelevant.

5. What size round, square, and device boxes are required to enclose six 8 AWG conductors to meet the minimum requirement?

The 4" x 2-1/8" round box (seven 8 AWG) is the only box of the 3 listed round boxes that meets the minimum requirement.

All of the square boxes listed can be used, however to meet the minimum requirement the 4" x 1¼" square box (six 8 AWG) is selected.

The 3" x 2" x 3-1/2" device box (six 8 AWG) is the only box listed that meet the minimum requirement.

6. What size junction box without the use of an extension ring is needed to enclose 18-14 AWG conductors for future use?

The only box listed in the table that can safely accommodate 18-14 AWG conductors is a 4-11/16" x 2-1/8" square box (42 in³). This size box has the capacity to hold up to 21-14 AWG conductors.

7. How many 8 AWG ZW conductors are permitted in a single gang FS box?

   a.   six   b. four   c. five   d. three

Four (4) conductors are permitted.

8. How many 6 AWG THW conductors are permitted in a 3¾" x 2" x 2½" masonry box?

Two (2) conductors are permitted.

When Box Contains Conductors of Different Sizes, Devices and Other Components

When a metal/nonmetal wiring enclosure such as an outlet box, device box, junction box, etc. is needed to enclose conductors of different sizes along with devices (switches, receptacles and other components (fixture studs or hickeys, manufactured cable clamps, etc.) NEC 314.16(B)(1) - (5) and Table 314.16(B) must be applied to determine the required box size. Small fittings such as locknuts, bushings, chase nipples, connectors, etc. are not required to be taken into consideration [See Figure 314.16(B)].

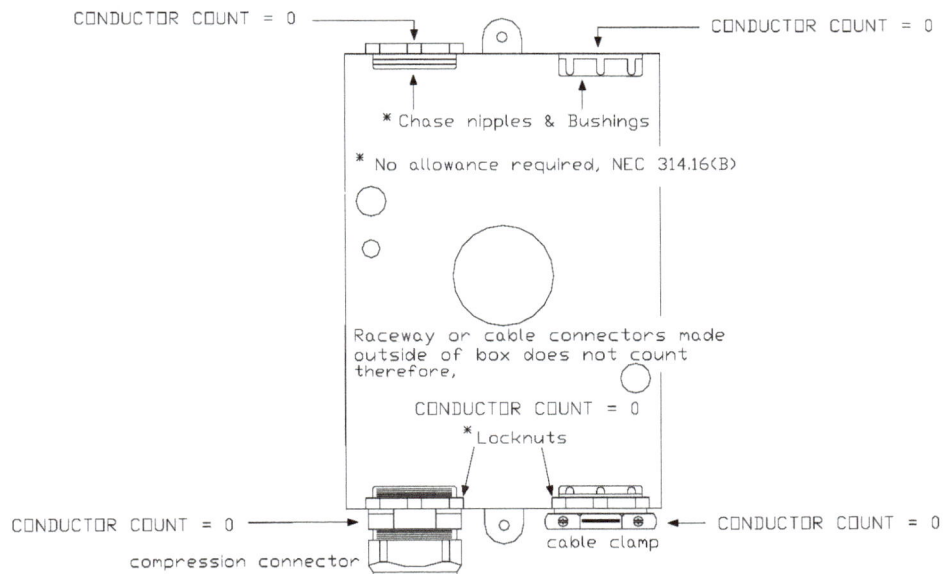

Figure 314.16(B) - Small Fittings

This consideration also *excludes* wirenuts or any other type splicing or grounding devices.

Refer to the following procedures,

1. **Conductor Fill** [NEC 314.16(B)(1)] [Refer to Figure 314.16(B)(1)]

   (a) For each conductor originating outside of the box and either terminates or is spliced inside the box the *conductor count* = **1**.

   (b) For each conductor that passes through the box without splice or termination the *conductor count* = **1**.

   (c) For each loop or coil of unbroken conductor 12" and greater (twice the minimum length required for free conductors per NEC 300.14) the *conductor count* = **2**. Where the loop or coil of an unbroken conductor is less than 12" the *conductor count* = **1.**

   (d) For those conductors that originates and terminates within the box such as pigtails and bonding jumpers the *conductor count* = **0**.

Figure 314.16(B)(1) - Conductor Fill

   (e) To determine the volume of those conductors described in **1(a)** through **(c)** above refer to Table 314.16(B).

   (f) *Exception* - (refer to Figure 314.16(B) [*Exception*]) *When one or more equipment grounding conductors and/or not over four (4) fixture wires smaller than 14 AWG enters a domed fixture or similar canopy and terminates within the box the *conductor count* = **0**.

   Simply said, "Equipment grounding conductors and up to four (4) 16 AWG and smaller fixture wires *can be omitted* from box fill calculations *if* they enter the box from a *domed luminaire* or *similar canopy* and terminate within that box."

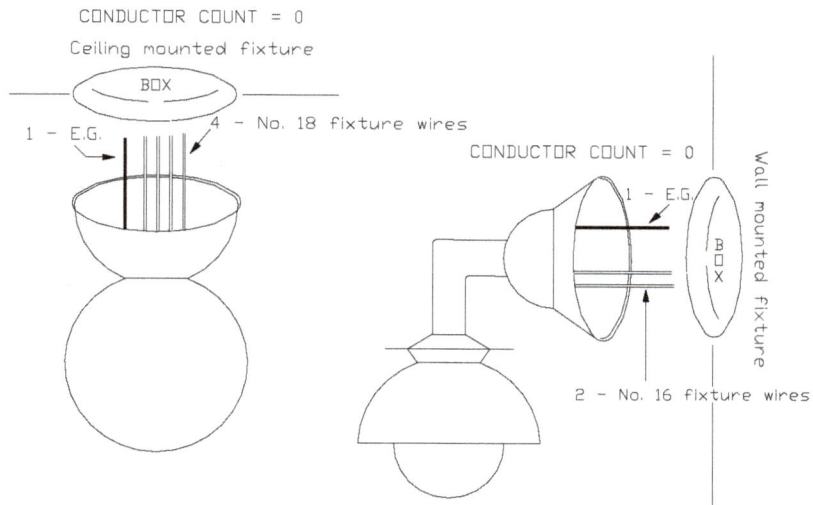

Figure 314.16(B) [*Exception*] - Domed Luminaires (Fixtures) or similar canopies

2. Clamp Fill [NEC 314.16(B)(2)] [Refer to Figure 314.16(B)(2)]

When a box contains one or more factory or field supplied internal cable clamps (whether used or unused) the *conductor count* = **1** and the volume of the conductor is based upon the largest conductor present in the box.

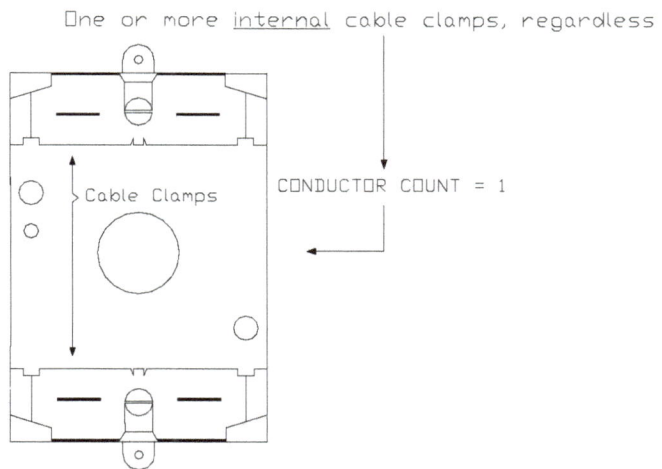

Figure 314.16(B)(2) - Clamp Fill

3. Support Fittings Fill [NEC 314.16(B)(3)][Refer to Figure 314.16(B)(3)]

When a box contains one or more fixture stud(s) or hickey(s) the *conductor count* = **1** per type fitting and the volume of the conductor is based upon the largest conductor present in the box.

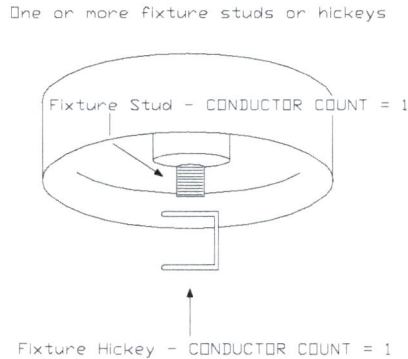

Figure 314.16(B)(3) - Support Fittings Fill

4. Device* or Equipment Fill [NEC 314.16(B)(4)][Refer to Figure 314.16(B)(4)(a) and (b)]

(a) When a single or duplex device is mounted on the same strap (or yoke) the *conductor count* = **2** and the volume of the conductor is based upon the largest conductor terminating on the device.

(b) When a device or utilization equipment is wider than any of the 2" single gang device* boxes described in Table 314.16(A) [Ex. 3"(L) x 2"(W) x 2½"(D)] the *conductor count* = **2** per box gang (required for device or utilization equipment mounting).

* refer to Article 100 for definition of device.

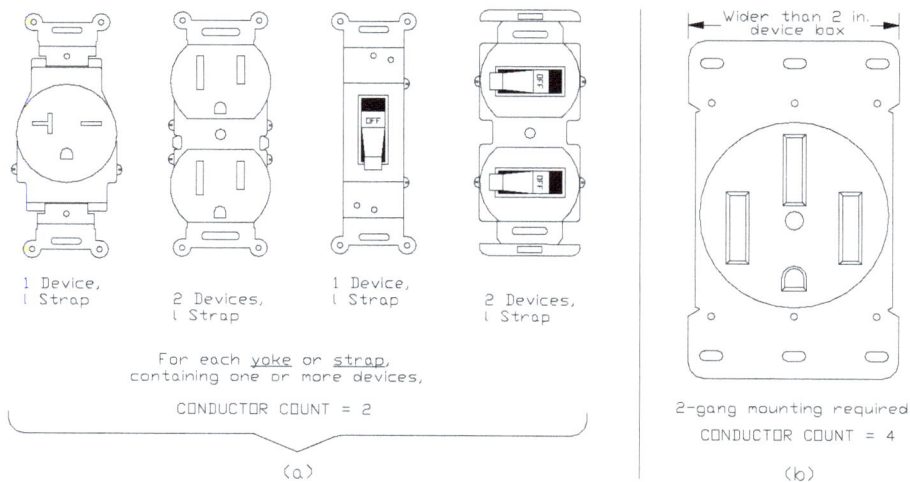

Figure 314.16(B)(4)(a) - Device or Equipment Fill

5. Equipment Grounding Conductor Fill [NEC 314.16(B)(5)][Refer to Figure 314.16(B)(5)]

(a) When one or more equipment grounding conductor(s) or equipment bonding jumper(s) enters a box the *conductor count* = **1** based upon the larger of the two present in the box.

(b) When an additional allowance is permitted per NEC 250.146(D) to include a separate set of equipment grounding conductors the *conductor count* = **1** [added with the conductor count in **(a)** above] based upon the largest equipment grounding conductor of the additional set.

NEC 250.146(D) - Isolated Receptacles - Where required to reduce electrical noise (electromagnetic interference) in the grounding circuit, a separate insulated grounding conductor may be run to the grounding terminal on an isolated grounding receptacle although this method will not satisfy the grounding requirements for the raceway system and outlet box.

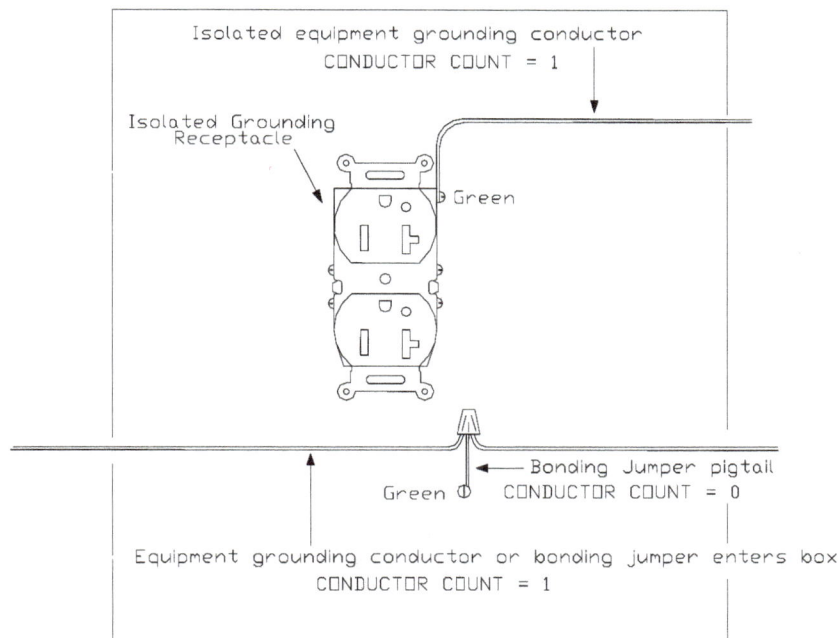

Figure 314.16(B)(5) - Equipment Grounding Conductor Fill

Table 314.16(B) - Volume Allowance Required per Conductor

9. What is the volume of a 10 AWG XHHW copper conductor according to Table 314.16(B)?

The volume of a 10 AWG conductor is 2.5 in³. The conductor volumes listed in Table 314.16(B) are not based on the type insulation covering the core conductor or the type material the insulation is made of.

314.16(B) - Box Fill Calculations (10. - 21.) [12]

10. List the conductor count per component(s) and procedure/NEC reference(s) as illustrated.

Components	Conductor Count	Procedure/NEC Reference(s)
1 - fixture stud	1	**3**
1 - duplex single pole switches	2	**4(a)**
1 - equipment bonding jumper (entering box)/ if smaller than equipment grounding conductor(s) also entering box)	1/0	**5(a)**
2 - cable clamps	1	**2**
1 - hickey	1	**3**
1 - neutral pigtail	0	**1(d)**
2 - conduit connectors with locknuts and bushings	0	**314.16(B)**
1 - single receptacle	2	**4(a)**
2 - transformer conductors (18 AWG) (doorbell chime)	2	**1(a)**
4 - grounding conductors **(if equipment bonding jumper(s) also enters box, yet smaller than grounding conductors)	1**	**5(a)**
1 - 13" unbroken conductor	2	**1(c)**
1 - 2-7/16" (width) range receptacle in 2-gang box	4	**4(b)**
3 - fixture wires with ground (16 AWG) ***(in domed fixture or similar canopy)	0***	**1(f)**
2 - conductors passing through	2	**1(b)**
2 - black and white conductor for light fixture	2	**1(a)**
2/5 - equipment grounds/18 AWG fixture wires (in canopy)	1	**1(f)**
1 - duplex receptacle	2	**4(a)**
6 - spliced conductors	12	**1(b)**

11. Two 12/2 Romex cables with grounds enter a box containing cable clamps from opposite sides. Both black wires of the cable terminate on a single-pole light switch. Determine the minimum size metal device box needed for this installation.

Determine conductor count

Conductors (black and white) = 4
Equipment ground = 1
Cable clamps = 1
Switch = 2
Total **8** (conductor count)

Because the conductors and equivalent components (clamps and switch) reference 12 AWG conductors, the device box can be sized according to Table 314.16(A) where a 3" x 2" x 3½" device box can be used to enclose **eight (8)** 12 AWG conductors or a combination of conductors and equivalent components.

12. Determine the minimum size octagonal metal box required when the box will contain a stud, a hickey, four cable clamps, a 14 AWG hot, neutral and equipment grounding conductor and two 18 AWG fixture wires with an equipment ground.

Determine conductor count

14 AWG (hot and neutral)	= 2
14 AWG equipment ground	= 1
Stud	= 1
Hickey	= 1
Cable clamps (4)	= 1
18 AWG fixture wires	= 0
18 AWG equipment ground	= 0
Total	**6** (conductor count)

A metal octagonal box can be size according to Table 314.16(A) where the minimum size box required for this installation is a 4" x 1¼" box based on **six (6)** 14 AWG conductors.

13. Determine the minimum size octagonal box required to house nine 14 AWG THWN copper conductors. Five of the conductors will pass straight through the box where one is coiled 14" while the remaining conductors will be spliced and pigtailed to a porcelain lampholder.

Determine conductor count

14 AWG conductors straight through	= 4
14 AWG conductor straight through coiled	= 2
14 AWG spliced conductors	= 4
14 AWG pigtail	= 0
	10

Per Table 314.16(A) a **4" x 2-1/8"** octagonal box is required for this installation.

14. Determine the total conductor count and the minimum size metal device box required to enclose the following components:

1 - 10 AWG hot conductor   1 - 10 AWG neutral conductor
1 - 10 AWG pigtail for neutral   1 - single receptacle (10 AWG terminating conductor)
4 - 14 AWG conductors passing through   1 - 10 AWG equipment bonding jumper

Question Nos. 14. and 15. will be used as examples to demonstrate the procedure for calculating box sizes per conductor count and volume.

Application

Step 1: List the quantity, size and type component.
Step 2: Determine each components conductor count per NEC references.
Step 3: List the volume of each conductor per Table 314.16(B).
Step 4: List the total conductor volume of all components.
Step 5: Compute the total volume of all conductors.

Step 1 Components	Step 2 Conductor Count	Step 3 Conductor Volume (in^3)	Step 4 Total Conductor Volume (in^3)
2-10 AWG conductors	2	2.50	5.0
1-10 AWG pigtail	0	0	0
1-10 AWG bonding jumper	1	2.50	2.5
4-14 AWG conductors	4	2.00	8.0
1-single receptacle (10 AWG)	2	2.50	5.0
TOTAL	9		**Step 5** 20.5

Step 6: Select an appropriate box according to the total volume of all conductors.

Results: Referring to Table 314.16(A), there is no minimum size device box to meet this application because all listed device boxes have a volume that is less than **20.5 in^3**. However, where applicable a smaller size device box could be used with either an extension ring, plaster ring or an industrial cover (depending on application) with a cubic capacity that will add to the volume of a smaller size device box that will either meet or exceed the **20.5 in^3** requirement.

15. What size square box is needed to enclose the following components?

    1 - 12 AWG hot conductor   1 - 12 AWG neutral conductor   1 - 14 AWG hot conductor
    1 - 14 AWG neutral conductor   1 - 12 AWG grounding conductor
  *1 - 14 AWG grounding conductor   2 - 14 AWG conductors passing through
   1 - duplex receptacle (12 AWG terminating conductor)
   1 - single receptacle-isolated grounding (14 AWG terminating conductor)

    *isolated equipment grounding conductor

Application

Step 1: List the quantity, size and type component.
Step 2: Determine each components conductor count per NEC references.
Step 3: List the volume of each conductor per Table 314.16(B).
Step 4: List the total conductor volume of all components.
Step 5: Compute the total volume of all conductors.

Step 1 Components	Step 2 Conductor Count	Step 3 Conductor Volume (in^3)	Step 4 Total Conductor Volume (in^3)
1-12 AWG hot conductor	1	2.25	2.25
1-12 AWG neutral conductor	1	2.25	2.25
1-12 AWG grd'g conductor	1	2.25	2.25
1-14 AWG hot conductor	1	2.00	2.00
1-14 AWG neutral conductor	1	2.00	2.00
1-14 AWG grd'g conductor*	1	2.00	2.00
2-14 AWG conductors passing through	2	2.00	4.00
1-duplex recep. (12 AWG)	2	2.25	4.50
1-single recep. (14 AWG)	2	2.00	4.00
TOTAL	12		Step 5 25.25

Step 6: Select an appropriate box according to the total volume of all conductors.

Results: Based on the total conductor volume, two of the square boxes listed in Table 314.16(A) cannot be used as is, whereas the other four remaining square boxes can. However, remember the question was, what size square box is needed, instead of asking for the minimum size. The square boxes sized smaller than **25.25 in^3** could be used, but only with an assembled section that will either make up the difference of the required volume or exceed it.

16. A two gang nonmetallic box is used to enclose a single pole light switch and a three-way light switch. Two 14/2 type NM-B cables enters the box. The two white conductors and the equipment grounding conductors are spliced together separately and a pigtail originating from the 14 AWG equipment grounding conductors terminate on the grounding terminal of the single pole switch. The two 14 AWG black conductors terminate on the brass terminals of the single pole switch. A 12/3 and a 12/2 type NM-B cables also enters the box. The black and red conductors of the 12/3 cable terminate on the traveler terminals of the three-way switch while the white conductors of both cables are spliced together. The remaining black conductor of the 12/2 cable terminates on the common terminal of the three-way switch. A 12 AWG pigtail originating from the spliced 12 AWG equipment grounding conductors terminates on the grounding terminal of the three-way switch. Determine the minimum required volume of the two gang nonmetallic box.

Because the box will contain conductors of different sizes along with devices, the equivalent volume per conductor and device must be determined based on the conductor volume allowance per Table 314.16(B).

Determine conductor count

12 AWG conductors	= 5
14 AWG conductors	= 4
12 AWG equipment ground	= 1

14 AWG equipment ground   = 0
Single pole light switch   = 2 (14 AWG - largest conductor)
Three-way switch   = 2 (12 AWG - largest conductor)

Totaling the 12 AWG and 14 AWG conductors together and calculating the total volume of each conductor yields,

$$8 - 12 \text{ AWG} = 8 \times 2.25 \text{ in}^3 = 18.00 \text{ in}^3$$
$$6 - 14 \text{ AWG} = 6 \times 2.00 \text{ in}^3 = \underline{12.00 \text{ in}^3}$$
$$\mathbf{30.00 \text{ in}^3}$$

The two gang nonmetallic plastic box must have a volume not less than **30.00 in³**.

17. A 4" square box is being used to install a duplex receptacle and a light switch. A black and white 12 AWG conductor will terminate on the receptacle and two blue 14 AWG conductors will terminate on the light switch. The conductors enter the box along with other conductors through a 1" EMT that has an attached grounding bushing. A 12 AWG bonding jumper is ran from the bushing to bond the box. Along with two cable clamps inside the box, a 12 AWG and 14 AWG equipment grounding conductor terminates on the devices while two 10 AWG and 8 AWG conductors pass through the box. A 3.5 in³ plaster ring will be mounted to the box to secure the receptacle and switch. Determine the minimum size square box needed.

Determine conductor count

14 AWG		12 AWG	
Blue conductors	= 2	Black and white conductors	= 2
Light switch	= 2	Receptacle	= 2
Equipment ground	= 0	Equipment ground	= 1
	4	Bonding jumper	= 0
			5
10 AWG		8 AWG	
Conductors passing thru	= 2	conductors passing thru	= 2
		cable clamps	= 1
			3

Consider total conductor volume per Table 314.16(B)

$$14 \text{ AWG} - 4 \times 2.00 \text{ in}^3 = 8.00 \text{ in}^3$$
$$12 \text{ AWG} - 5 \times 2.25 \text{ in}^3 = 11.25 \text{ in}^3$$
$$10 \text{ AWG} - 2 \times 2.50 \text{ in}^3 = 5.00 \text{ in}^3$$
$$8 \text{ AWG} - 3 \times 3.00 \text{ in}^3 = \underline{9.00 \text{ in}^3}$$
$$\mathbf{33.25 \text{ in}^3}$$

According to NEC 314.16(A) the total volume of a box shall include the volume of an assembled section, such as the 3.5 in³ plaster ring that will be mounted to the box. Because of this, the minimum size square box that can be used is a 4" x 2-1/8" box which has a minimum cubic inch capacity of **30.3 in^3**. When the volume of the box and plaster ring are added together the total **(33.8 in^3)** will exceed the total volume calculated **(33.25 in^3)** per Table 314.16(B).

18. An isolated receptacle is installed in a metallic device box. A 12 AWG hot and grounded conductor terminates on the receptacle along with a 12 AWG isolated equipment grounding conductor. The box is grounded with a separate 10 AWG equipment grounding conductor. Determine the minimum size device box needed for this installation.

Determine conductor count

12 AWG (hot and grounded conductors)	= 2
12 AWG (isolated equipment ground)	= 1
Receptacle (12 AWG)	= 2
10 AWG equipment ground	= 1

$$12 \text{ AWG - 5 x 2.25 in}^3 = 11.25 \text{ in}^3$$
$$10 \text{ AWG - 1 x 2.50 in}^3 = \underline{2.50 \text{ in}^3}$$
$$\mathbf{13.75 \text{ in}^3}$$

The minimum size metal device box that can be used is a **3" x 2" x 2¾"** box.

19. A 4-11/16" x 2-1/8" square box contains six 10 AWG THWN copper conductors. How many 6 AWG THWN aluminum conductors can be pulled through the box without exceeding the box's cubic inch capacity?

To get started, let's examine Table 314.16(A) to determine the minimum volume (cubic capacity) for the 4-11/16" x 2-1/8" square box. The listed minimum volume for this box is 42 in^3. Now Table 314.16(B) is referenced to determine the volume for a 10 AWG and 6 AWG conductor.

The only data listed in Table 314.16(B) is the conductor's volume; the type material the conductor is made of is irrelevant. The listed volume for a 10 AWG and a 6 AWG conductor are 2.5 in^3 and 5.0 in^3, respectively.

The number of 10 AWG conductors contained in the square box is multiplied by the conductor's volume to obtain the total volume.
$$6 \text{ x } 2.5 \text{ in}^3 = 15 \text{ in}^3$$

The total volume of the 10 AWG conductors is then subtracted from the volume of the square box.
$$42 \text{ in}^3 - 15 \text{ in}^3 = 27 \text{ in}^3$$

The remainder is then divided by the volume of the 6 AWG conductors to determine the number of 6 AWG conductors that can be pull through the square box.

$$27 \text{ in}^3/5.0 \text{ in}^3 = \mathbf{5.4 \text{ conductors}}$$

The results, 5.4 have to be rounded down to 5. Rounding the results up to 6 would exceed the maximum number of conductors allowed in the 42 in^3 square box. Therefore, the number of 6 AWG conductors that can be added to the square box is **5**.

20. How many 14 AWG THW conductors can be pulled through a 4-11/16" x 1½" square box, if the box contains two receptacles where a 14 AWG ungrounded and grounded conductor along with one equipment grounding conductor each terminates on the receptacles?

Determine conductor count

14 AWG conductors	= 4
14 AWG equipment grounds	= 1
Receptacles (2)	= 4
	9

Actually, the answer can be derived by either subtracting the total conductor count (nine) from the maximum number of conductors (14 - 14 AWG) listed in Table 314.16(A) for a 4-11/16" x 1½" square box which leaves **five** 14 AWG conductors that can be added to the square box.

Another way to obtain the answer is to first derive the total conductor volume based on **nine (9)** 14 AWG conductors, (9 x 2.00 in^3 = 18 in^3) then derive the difference between the box's minimum volume (29.5 in^3) and the total conductor volume (18 in^3),

$$29.5 \text{ in}^3 - 18 \text{ in}^3 = 11.5 \text{ in}^3 \text{ (remainder)}$$

and finally, determine the number of 14 AWG conductors that can be added by dividing the volume allowance of a 14 AWG conductor (2 in^3) into the remainder,

$$11.5 \text{ in}^3 / 2 \text{ in}^3 = 5.75 \text{ (rounded down) to } \mathbf{5}$$

Both methods provide the same answer, **five** 14 AWG conductors can be added to the square box.

21. A 8/3 Romex cable with 10 AWG equipment ground is being used to supply an electric dryer receptacle that has a width measuring approximately 2½". What size 2-gang plastic box is needed to enclose the cable and receptacle?

Considering the Romex cable where three 8 AWG conductors and a 10 AWG bare conductor are present the conductor's count and total volume amounts to,

8 AWG - 3 x 3 in^3	=	9.0 in^3
10 AWG - 1 x 2.5 in^3	=	2.5 in^3
		11.5 in^3

Now considering the dryer receptacle which requires a double volume allowance and the consideration of the 2-gang box for mounting the receptacle, the equivalent conductor count and volume amounts to,

Dryer receptacle wider than 2"
2-gang box - double allowance **(2)** per gang **(2)** [2 x 2 = 4]
Total conductor count = **4**

Based on an 8 AWG conductor being the largest conductor connected to the receptacle per NEC 314.16(B)(4) and the total conductor count @ 3 in^3 per conductor, the total volume equals **12 in^3** (4 x 3 in^3).

When added to the Romex cable total calculated conductor volume (12 in^3 + 11.5 in^3), the 2-gang plastic box must have a cubic capacity no less than **23.5 in^3**.

CONDUIT BODIES FILL CALCULATIONS

The same provisions outlined in NEC 314.16(B) are used along with NEC 314.16(C) to determine the maximum number of conductors permitted in a conduit body. This data is then used to properly size an appropriate conduit body.

Conduit bodies, other than short radius enclosures, containing 6 AWG conductors or smaller cannot be used with a raceway that has a larger cross-sectional area than the conduit body according to NEC 314.16(C)(1). Further provisions are extended in NEC 314.16(C)(2) which requires the volume (cubic capacity) of a conduit body to be displayed when such enclosure will contain devices, splice and tap conductors. In order to clearly understand both NEC requirements the following means are provided.

RELATING to **NEC 314.16(C)(1)**

The cross-sectional area [see Figure 314.16(C)(1)] of a conduit body must not be less than twice the cross-sectional area of the raceway to which it is attached.

```
Conduit Body "E" has one raceway entry
Conduit Bodies "C", "LB", "LL", & "LR" has two raceway entries
Conduit Body "T" has three raceway entries
Conduit Body "X [+]" has four raceway entries
```

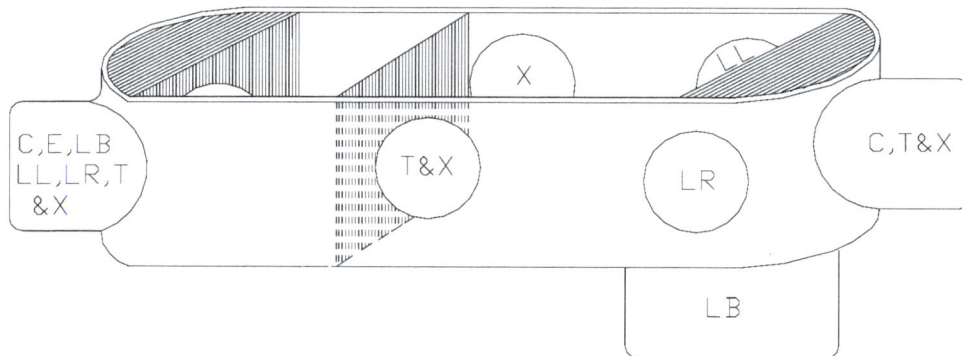

Figure 314.16(C)(1) - Illustrated Display of Various Conduit Bodies

Indeed, the design and construction of conduit bodies are the exclusive responsibility of its manufacturer. However, there is also a responsibility of the installer to ensure that such enclosures are installed in compliance with the NEC. NEC 110.3(B) states that listed or labeled equipment must be installed and used in accordance with any instructions included with equipment. Although, the cross-sectional area of a conduit body is not provided, manufacturers instead specify the maximum number and size of conductors a conduit body is allowed to contain which must be adhered to.

Typically, conduit bodies of sizes ¾" and larger list the maximum number of conductors per conductor size allowed in a conduit body along with the volume of the enclosure. Because the volume is listed in conduit bodies, this information can be used to ensure NEC 314.16(B) is complied with when 6 AWG or smaller conductors are enclosed.

The maximum number of conductors permitted to be enclosed in a conduit body shall be based upon **Table 1** of **Chapter 9**. The provisions of NEC 314.16(C)(1) are applied when an installation will not include those components described in NEC 314.16(C)(2).

314.16(C)(1) - General (Conduit Bodies) (22. - 23.) [2]

22. A type LB conduit body is needed to route seven 8 AWG RHH conductors. If the conductors are enclosed in an attached 1¼" EMT, what size LB conduit body is required for this installation?

Per **Table 4** of **Chapter 9** the maximum number of conductors permitted for this installation must be based on the 40 percent fill capacity according to NEC 314.16C(1). **Table 4** list the cross-sectional area of a 1¼" EMT based on a 40 percent fill capacity at .598 in^2. Because the cross-sectional area of the conduit body is required to be twice that of the 1¼" EMT, the end results amounts to,

$$.598 \text{ in}^2 \times 2 = 1.1960 \text{ in}^2$$

Given that conduit bodies are not marked with a cross-sectional area, the following approach is taken to determine the minimum size LB required. Again, referring to **Table 4** of **Chapter 9**, the next size EMT that reflects a cross-sectional area larger than 1.1960 in^2 yields a 2" trade size which has a cross-sectional area of 1.342 in^2 (40 percent fill capacity). Because a 2" EMT has a cross-sectional area larger than 1.1960 in^2, a 2" LB conduit body or larger is required for this installation. For such an installation a reducing bushing would be also required.

23. What size LB conduit body would be required if the conductors in question No. 22. were now four times as many and enclosed in an attached 2½" EMT?

Table 4 list the cross-sectional area of a 2½" EMT based on a 40 percent fill capacity at 2.343 in^2. Again, because the cross-sectional area of the conduit body is required to be twice that of the 2½" EMT, the end results amounts to,

$$2.343 \text{ in}^2 \times 2 = 4.686 \text{ in}^2$$

Per **Table 4** of **Chapter 9**, the next size EMT that reflects a cross-sectional area larger than 4.686 in^2 yields a 4" trade size which has a cross-sectional area of 5.901 in^2 (40 percent fill capacity). Because a 4" EMT has a cross-sectional area larger than 4.686 in^2, a 4" LB conduit body or larger is required for this installation.

As it pertains to question Nos. 22. and 23., refer to **Table C.1** of **Informative Annex C**. Also, for both installations a reducing bushing would be needed to make the required attachments.

RELATING to NEC 314.16(C)(2)

Where *splices*, *taps* or *devices* will be enclosed in a conduit body, only those conduit bodies that are durably and legibly marked by the manufacturer with their volume are permitted to be used.

The maximum number of conductors permitted to be enclosed in the conduit body shall be based upon NEC and Table 314.16(B). The total volume of the maximum number of conductors shall not exceed the listed volume of the conduit body.

314.16(C)(2) - With Splices, Taps, or Devices (Conduit Bodies) (24. - 28.) [5]

24. An unmarked ¾" type E conduit body appears to be large enough to install a single pole switch and four 14 AWG conductors. Two of the 14 AWG conductors will terminate on the switch while the remaining two conductors will be spliced. Can this conduit body be used for such installation?

According to NEC 314.16(C)(2), if the cubic capacity (volume) of a conduit body is not marked in the enclosure, *splices*, *taps* or *devices* are not permitted to be installed. Therefore, this enclosure cannot be used for such installation.

25. How many 10 AWG THWN-2 copper conductors can be spliced inside a type C conduit body if the enclosure has a marked cubic capacity of 24 in^3?

According to NEC 314.16(C)(2), a conduit body can contain spliced conductors if the volume is durably and legibly marked by the manufacturer on the enclosure. Referring to Table 314.16(B), a 10 AWG conductor has a volume of 2.5 in^3. Therefore,

$$24 \text{ in}^3 / 2.5 \text{ in}^3 = 9.6 \text{ (rounded down) to } 9$$

Any splicing combination can be used as long as it doesn't involve more than nine 10 AWG spliced conductors.

Although not mentioned, correction (ambient temperature) and adjustment factors must be considered in all similar installations to ensure adequate conductor ampacity.

26. A Type C aluminum conduit body will contain the following:

    1 - single pole switch
    2 - 14 AWG conductors passing through
    2 - 10 AWG conductors terminating on the switch
    2 - 10 AWG spliced conductors
    2 - 10 AWG equipment grounding conductors

Determine the required volume of the conduit body.

The same procedure used with questions Nos. 14. and 15. will be used to calculate the required volume of a conduit body per conductor count and volume.

Application

Step 1: List the quantity, size and type component.
Step 2: Determine each components conductor count per NEC references.
Step 3: List the volume of each conductor per Table 314.16(B).
Step 4: List the total conductor volume of all components.
Step 5: Compute the total volume of all conductors.

Step 1 Components	Step 2 Conductor Count	Step 3 Conductor Volume (in^3)	Step 4 Total Conductor Volume (in^3)
2 - 14 AWG conductors	2	2.00	4.00
8 - 10 AWG conductors (switch, terminating,			
spliced and grounding)	7	2.50	17.50
	9		Step 5 21.50

Step 6: Select an appropriate conduit body according to the total volume of all conductors.

Results: The required volume of the conduit body must be no less than 21.50 in^3.

27. Where marked, determine the minimum size conduit body needed for the installation in question No. 24.

Per NEC 314.16(B)(1) and (4), since 14 AWG conductors are terminated on the switch, the switch will count as two 14 AWG conductors while the two terminating conductors and the two spliced conductors will total four additional 14 AWG conductors resulting in an overall total of six 14 AWG conductors. Therefore, six 14 AWG conductors at 2 in^3 each per Table 314.16(B) yields,

$$6 \times 2 \text{ in}^3 = 12 \text{ in}^3$$

As a minimum, the ¾" Type E conduit body must be marked no less than 12 in^3.

28. How many spliced 6 AWG THWN conductors are permitted to be enclosed in a 1" PVC conduit body having a cubic capacity of 25 in^3?

According to NEC 314.16(B)(1) a spliced conductor is counted once and the volume of a 6 AWG conductor per Table 314.16(B) regardless of insulation type is 5 in^3. Therefore, to determine the number of spliced conductors permitted in the enclosure the volume of the conductor is divided into the volume of the conduit body.

Therefore,

$$25 \text{ in}^3/5 \text{ in}^3 = 5$$

As a result, **five** 6 AWG spliced conductors are permitted to be enclosed in the 1" PVC conduit body.

THE PULL BOX

The flexibility of smaller size conductors makes the job of wire pulling much easier compared to larger size conductors. Six AWG or smaller conductors that are required to be pulled at lengths of great distance can utilize the same type metal boxes as earlier discussed or similar type nonmetallic boxes as pull boxes also.

For larger size conductors the job of wire pulling becomes more troublesome and difficult. When considering the distance of a run and the number of bends an installation may require, it becomes necessary to install pull boxes to help overcome and reduce wire pulling difficulties. In most installations one box can be used for a number of raceway runs. Pull boxes are also used when large conductors are required to be spliced.

The NEC reference three types of pull boxes; the *straight pull box*, the *angle pull box* and the *U pull box*. When raceway contains single conductors of sizes 4 AWG or larger or multi-conductor cable containing conductors of similar sizes the dimensions of a pull box must be sized according to NEC 314.28.

Regardless of whether a pull box contains large or small size conductors, the box must be installed where the wiring installed inside the box is accessible without having to excavate or remove any parts of a permanent structure.

pull box - A box so located to ease the task of pulling conductors through raceway(s) from one location to the next.

ACCORDING TO THE NEC

SIZING PULL AND JUNCTION BOXES

The following NEC sections and tables are applicable for understanding how pull and junction boxes along with conduit bodies are sized and selected per Articles 314 and 312.

NEC 314.28 - Pull and Junction Boxes and Conduit Bodies - NEC 314.28(A) - (E) must be complied with when boxes and conduit bodies are used for pull or junction boxes.

Exception - Terminals housings supplied with motors shall comply with NEC 430.12.

NEC 314.28(A) - Minimum Size - When raceways contain single conductors that are 4 AWG or larger or when *cables contain conductors that are 4 AWG or larger, the minimum dimensions of pull or junction boxes and conduit bodies that are used in conjunction with these raceways or cables shall adhere to the following:

*When multi-conductor cable is exclusively used, the largest raceway requirement is determined by using the minimum size raceway that would be used to enclose single conductors of the same quantity and size similar to those enclosed in the multi-conductor cable.

(1) Straight Pulls - When a box (or conduit body-Types C, T and X [+]) is needed to make a straight pull, the length of the box shall be determined by the size of the raceway being used. The length of the box must not be less than *eight times* the trade size of the largest raceway. Types C, T and X are the only conduit bodies applicable to this section.

(2) Angle or U Pulls - When a box (or **conduit body) is needed to make an angle or U-pull, the distance between the point where the raceway enters the box and the opposite wall shall not be less than *6 times* the trade size diameter of the largest raceway. If more than one raceway entry is made (into the box), the distance shall be increased by adding to the diameters of each raceway that enters the box in one row on the same wall. Each row must be calculated individually. The single row that provides the maximum distance shall be used.

** used only for angle pulls, applies to Types LB, LL, LR, T and X conduit bodies.

Raceway entries containing the same conductor shall be placed at a distance not less than *6 times* the trade size diameter of the largest raceway.

Exception - When a raceway or cable enters the wall of a box or conduit body opposite the removable cover of the box or conduit body, the distance from that wall to the removable cover shall not be less than the distance listed in Table 312.6(A) for one wire per terminal. Type LB is the only conduit body applicable to this *Exception*.

(3) Smaller Dimensions - Boxes or conduit bodies of lesser dimensions than those required in NEC 314.28(A)(1) and (2) may be used to install combinations of conductors that are less than the maximum conduit or tubing fill permitted in **Table 1** of **Chapter 9** provided the boxes or conduit bodies are approved for and are permanently marked with the maximum number and maximum size of conductors allowed.

Table 312.6(A) - Minimum Wire-Bending Space at Terminals and Minimum Width of Wiring Gutters (Sizing Depth of a Pull Box/Conduit Body) - Although Table 312.6(A) is used to provide the standard for deflecting conductors in cabinets or cutout boxes, it is also used to determine the (depth) of pull boxes and type LB conduit bodies with removable covers.

APPLYING THE NEC

To establish a clear interpretation according to the NEC of how boxes are sized to accommodate *straight*, *angle* or *U pulls*, each type installation will be demonstrated and discussed individually. It is always best to sketch the configuration of the pull box according to desired installation before performing any calculations. So often by doing this, guesswork is reduced, workmanship is increased and most importantly, costly field errors can possibly be prevented.

THE STRAIGHT PULL

- **As defined**

  A box or conduit body used to conduct a straight pull where similar raceway or a group of raceways are mounted on opposite walls (hubs) of the enclosure to ease the transition of conductors being routed from an initial starting point to a final destination.

- **As sized**

  When sizing a box to conduct a straight pull [See Figure 314.28(A)(1)] NEC 314.28(A)(1) only requires,

  The length (**L**) of the box (or conduit body) to be *eight times* the trade size of the largest raceway, where

  $$L = 8 \text{ x trade size (metric designator) of largest raceway}$$

Note: Eight times the trade size of the largest raceway is only an NEC minimum requirement, if needed the multiplier can be exceeded.

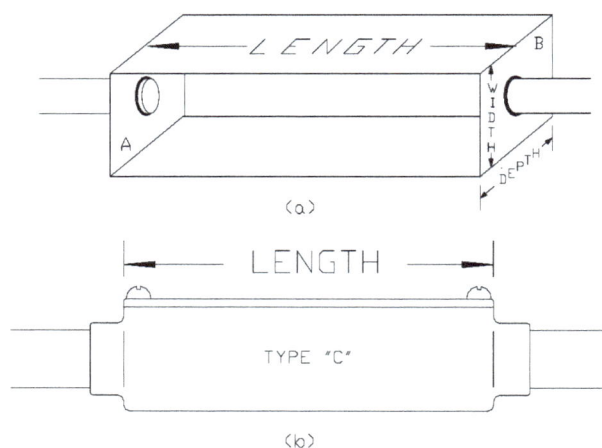

FOR STRAIGHT PULL BOX (a) and CONDUIT BODY (b)
Length (L) must be 8 times the diameter of largest raceway

Figure 314.28(A)(1) - Straight Pull Box in 3 - Dimensions and Applicable Conduit Body

Although the NEC only specifies the length (**L**) of a straight pull box, the width (**W**) and the depth (**D**) of the box should also be considered before the actual size of the box is determined. As a result, consider the following *alternatives* (non-NEC directives) but remember, they are just that, *alternatives* that are non-binding. Because the length of a pull box is based on the distance between opposite walls which raceway is mounted, the width and depth as shown in Figure 314.28(A)(1) must be altered when the raceway is vertically mounted.

ALTERNATIVES

Refer to Figure 314.28(A)(1).1. The *width* and *depth* of the box should be wide and deep enough to accommodate,

A - spacing from the outer edges of the box to the nearest locknut(s),
B - spacing between locknuts (when applicable), *and*
C - the diameter(s)*of the largest locknut or the largest locknut combinations per column(s) and row(s),

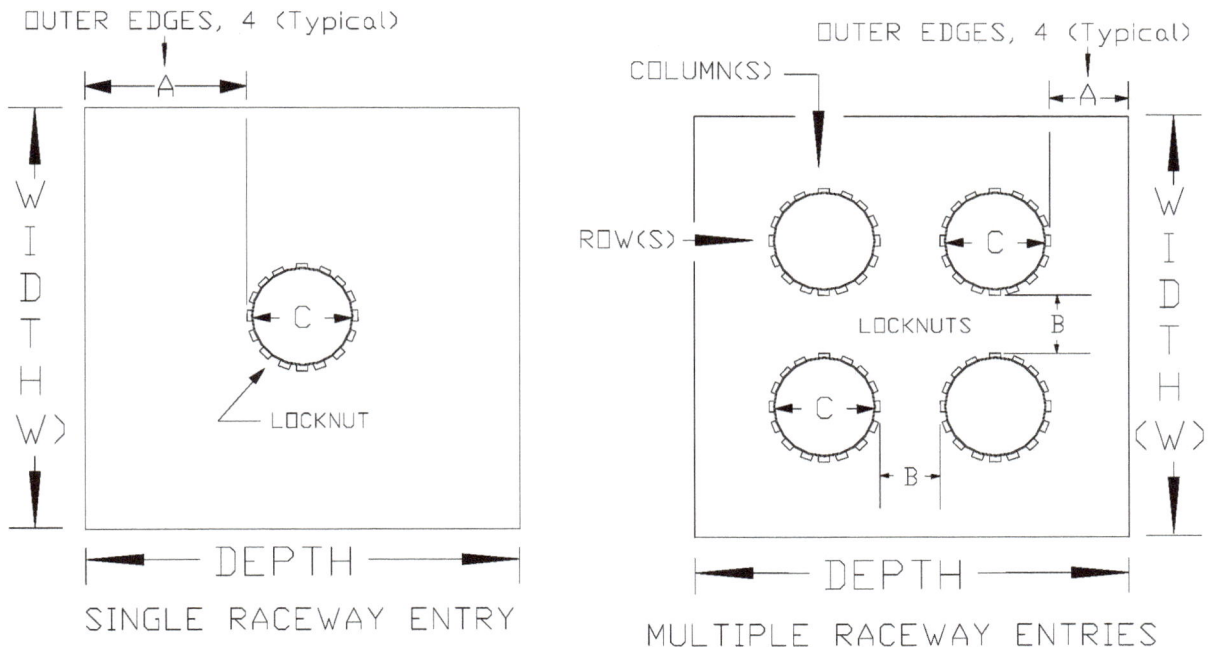

Figure 314.28(A)(1).1 - Walls of Straight Pull Box with single and multiple raceway entries

*Because locknuts are generally equal in diameter to bushings, yet have a larger diameter than raceway and other type fittings, the dimensions (diameter) of the locknuts are used instead. As a rule of thumb, use a minimum of 2" spacing from the outer edges of the box and between each locknut. However, just remember, this is only a rule of thumb, spacing is discretionary. See **Locknut Dimensions Table**.

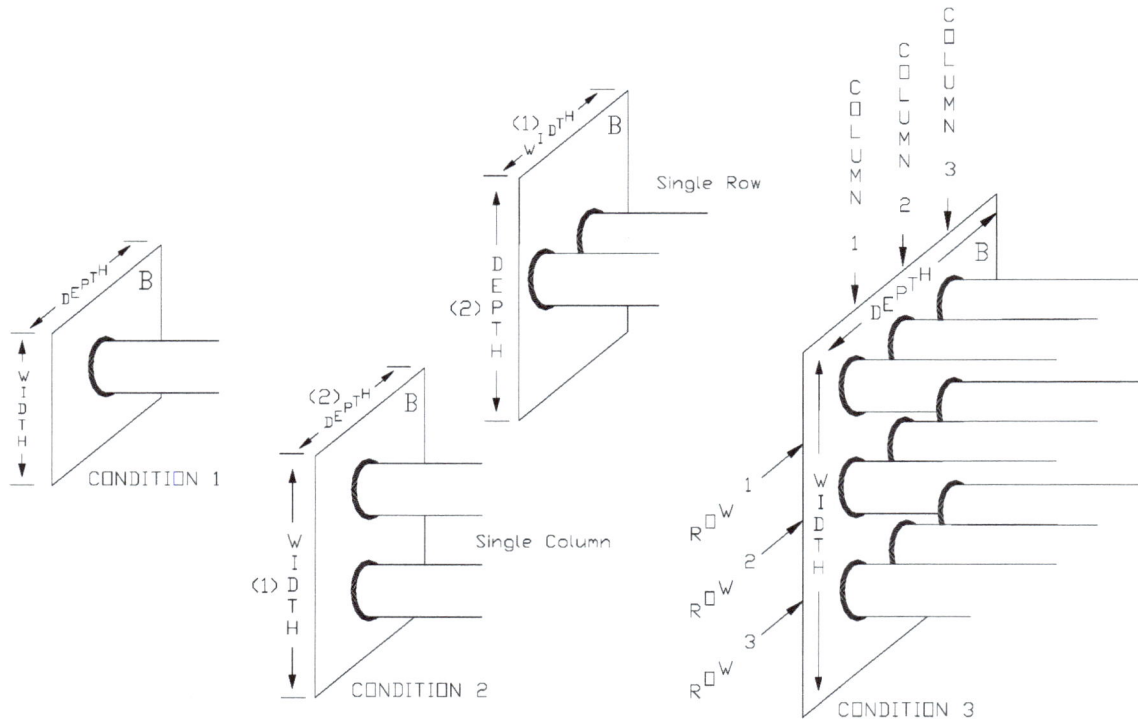

Figure 314.28(A)(1).2 - Conditional representation of Straight Pull Box

Figure 314.28(A)(1).2 provides a sectional representation of both walls of the pull box shown in Figure 314.28(A)(1)(a). Since both walls are similar in dimensions, only wall **B** is displayed. Figure 314.28(A)(1).2 displays the 3 conditions primarily to enhance the comprehension of the above *alternatives* for determining the **minimum** width and depth of a straight pull box. The conditions are as follows,

Condition 1

When walls **A** and **B** contain only *one* raceway entry, the wall containing the largest locknut will be used to determine the width (W) and depth (D) of the walls of the box where,

W and **D** = 4" + diameter of largest locknut

Condition 2

When walls **A** or **B** contain *two* raceway entries that form a single column or row,

(1) - the width or depth of the box is determined by the wall containing the largest locknut combinations resulting from the addition of each individual locknut's diameter where,

W or **D** = 4" + (2" x number of spaces between locknuts) + (diameters of largest locknut combinations)

(2) - the depth or width of the box is based on the largest locknut of the combination as derived in **Condition 1**.

Condition 3

When walls **A** or **B** contain raceway entries that form *multiple* columns and rows, the width and depth of the box is determined by the wall containing the largest locknut combinations per row and/or column and adding the resulting combinations where,

W and **D** = 4" + (2" x number of spaces between locknuts) + (diameters of largest locknut combinations per column and/or row)

The above conditions can also be used for determining the depth of an Angle and U-Pull Box. If U-pull Box contains raceway entry opposite the box cover, these conditions cannot be used. Again, all calculated values derived from **Conditions 1-3** should always be recognized as *minimum* values and rounded off to the nearest whole number.

Locknut Dimensions Table
(Approximate)

Conduit Size	Locknut (Outer Diameter-inches) Fractional	Decimal
3/8"	1-1/16	1.0625
½ "	1-3/32	1.0938
¾ "	1-11/32	1.3438
1"	1-21/32	1.6563
1¼ "	2-1/8	2.1050
1½ "	2-3/8	2.3750
2"	2-31/32	2.9688
2½ "	3-17/32	3.5313
3"	4-5/16	4.3105
3½ "	5	5.0000
4"	5-3/8	5.3750
4½ "	6-3/32	6.0938
5"	6-3/4	6.7500
6"	7-19/32	7.5938

314.28(A)(1) - Straight Pulls (Pull and Junction Boxes and Conduit Bodies) (29. - 36.) [7]

29. A box being used for a straight pull has 3" conduit mounted on opposite sides of the box. Determine the minimum length of the box.

As mentioned earlier, it is always best to sketch the configuration of the desired pull box installation before performing any calculations, so the first approach should always be to *"sketch the layout of the box"*.

The minimum length of the box is determined by multiplying the size of the largest conduit eight times. Since the only size conduit being used is 3" and is therefore the largest raceway, the minimum length equals, 8 x 3" = 24".

The minimum length of the box must be 24" long.

30. It has been determined that a straight pull box will contain different runs of raceways. One wall of the box will contain a 1", 2½", ½", ¾" and 2" raceway, while the opposite wall of the box will contain a ¾", 2", 1¼", 1" and 1½" raceway. Determine the minimum length of the pull box.

"Sketch the layout of the box".

Although the question does not provide a specific format pertaining to the configurations of the raceways, it's really not required since the question only asked to determine the minimum length of the box which is determined based on the largest raceway.

Since the 2½" raceway is the largest of the combination, the minimum length of the straight pull box equals, 8 x 2½" = 20".

The minimum length of the straight pull box must be 20" in length however; any box exceeding the minimum length can be used if desired.

31. Determine the minimum overall dimensions of a straight pull box that has a ½", 1", 2" and a 3" conduit horizontally aligned in one row on opposite walls of the box.

"Sketch the layout of the box".

Again, the minimum length (**L**) of the box is determined by multiplying the size of the largest conduit *eight times*. Therefore, since the 3" conduit is the largest of the conduits, the minimum length of the straight pull box equals, 8 x 3" = 24".

APPLYING THE ALTERNATIVES

The *width* and *depth* of the box are determined based on the given *alternatives*, (refer to Figure 314.28(A)(1).2, if needed).
According to **Condition 2(1)**

The *alternative* width (W) is determined by the wall containing the largest locknut combinations. Since both walls will contain similar conduit entries, only one calculation is required to determine the width. The outer diameter of a ½", 1", 2" and 3" locknut is 1.0938", 1.6563", 2.9688" and 4.3105" respectively, 10.0294" combined.

Therefore,

$$\mathbf{W} = 4" + (2" \times 3) + 10.0294" = 20.0294" (20")$$

According to **Condition 2(2)**

The *alternative* depth (D) based on the largest locknut of the combination is,

$$\mathbf{D} = 4" + 4.3105" = 8.3105" \ (8")$$

The minimum overall dimensions are,

$$(\text{per NEC}) \ \mathbf{L} = 24" \ (\text{per } \textit{alternative}) \ \mathbf{W} = 20" \text{ and } \mathbf{D} = 8"$$

Use a standard size pull box based on the minimum length and if desired the *alternative* width and depth dimensions.

32. The horizontal walls of a straight pull box will contain the following raceway entries,

Row 1 -	2",	2½ ",	3",	4"
Row 2 -	¾",	2",	1¼ ",	2"
Row 3 -	1½ ",	3",	4",	2"

Determine the minimum length of the box and the *alternative* width and depth.

"Sketch the layout of the box".

Based on the largest raceway entry which is the 4" raceway the minimum length of the box is calculated, 8 x 4" = 32".

Now closely observe the layout of the box. Using **Condition 3**, the *alternative* width and depth of the box is determined per column and/or row using the formula,

$$\mathbf{W} \text{ and } \mathbf{D} = 4" + (2" \text{ x number of spaces between locknuts}) + (\text{diameters of}$$
$$\text{largest locknut combinations per column and/or row})$$

Per Column (results rounded-off)

$$W(1) \ 4" + (2" \text{ x } 2) + 2.9688"(2") + 1.3438"(\ ¾") + 2.3750"(1½ ") = 14.69"$$
$$W(2) \ 4" + (2" \text{ x } 2) + 3.5313"(2½") + 2.9688"(2") + \ 4.3105"(3") = 18.81"$$
$$W(3) \ 4" + (2" \text{ x } 2) + \ 4.3105"(3") + 2.1050"(1¼") + 5.3750"(4") = 19.79"$$
$$W(4) \ 4" + (2" \text{ x } 2) + \ 5.3750"(4") + 2.9688"(2") + \ 2.9688"(2") = 19.31"$$

$$19.79" \ (20")$$

Per Row (results rounded-off)

D(1) 4" + (2" x 2) + 2.9688"(2") + 3.5313"(2½") + 4.3105"(3") + 5.3750"(4") = 24.19"
D(2) 4" + (2" x 2) + 1.3438"(¾") + 2.9688"(2") + 2.1050"(1¼") +2.9688"(2") = 17.39"
D(3) 4" + (2" x 2) + 2.3750"(1½") + 4.3105"(3") + 5.3750"(4") + 2.9688"(2") = 23.03"

24.19" (24")

Condition 3 states that the *width* and *depth* of a straight pull box is determined by the wall containing the largest locknut combinations per row and/or column. In this situation, both walls contain the same raceway entries. As a result, the calculations derived from *column 3* and *row 1* will be used to determine the *alternative* width and depth of the pull box or the largest calculation between the two can be used.

The minimum length and the *alternative* width and depth of the pull box is,

(per NEC) **L** = 32" (per *alternative*) **W** = 20" (or 24") and **D** = 24"

Use a standard size pull box based on the minimum length and if desired the *alternative* width and depth dimensions.

33. What size pull box is required when two 3" conduits are mounted on opposite walls of the box containing 3/0 AWG conductors?

The minimum length of the box is based on the 3" raceway where, 3" x 8 = 24".

The minimum length of the pull box must be at least 24".

34. A 5" raceway enters the right wall of a straight pull box while two 2" raceways enter the left wall of the box. The raceways will enclose conductors ranging from 1/0 AWG up to 250 kcmil. Determine the minimum length of box.

The minimum length of the box is based on the 5" raceway where, 5" x 8 = 40".

The minimum length of the pull box must be at least 40".

35. A Type C condulet containing 4-350 THWN conductors is required for a straight pull. Three and a half-inch (3½") raceway are installed in both hubs of the enclosure. Determine the minimum length of the condulet. Refer to Figure 314.28(A)(1)(b).

The minimum length of the condulet is determined by multiplying the size of the largest raceway being used eight times. Therefore, the minimum length of the condulet equals, 8 x 3½" = 28".

The minimum and only requirement per NEC is that the length of the conduit body must be 28" long.

36. A Type "T" conduit body is used to pull 4 - 4/0 AWG THWN copper conductors through 2" raceway mounted in common hubs of the enclosure. Determine the minimum length of the conduit body.

Because this enclosure is being used for the purpose of making a straight pull, the minimum length of the conduit body must be 16" (2" x 8).

THE ANGLE PULL

- ### As defined

    A box or conduit body used to transition the routing of conductors by means of a single or group of raceway(s) of similar or dissimilar sizes which are mounted on adjacent walls (hubs) of the box (conduit body).

- ### As sized

    When sizing a box or conduit body to conduct an angle pull [See Figure 314.28(A)(2)] NEC 314.28(A)(2) requires,

    (1) The length (**L**) and width (**W**) distances between each point where raceway enters the inside of the box or conduit body (length only) and the opposite wall must not be less than *six times* the trade size of the largest raceway, where

    L and **W** = 6 x trade size of largest raceway

    and

    (2) The length and width distances shall be increased for additional entries by the amount of the sum of the diameters of all other raceway entries in the same row on the same wall of the box, where

    L and **W** = 6 x trade size of largest raceway plus diameters of other raceway

    (3) Each row much be calculated individually

    and

    the single row that provides the maximum distance shall be used.

    (4) The diagonal distance (**DD**) between raceway entries enclosing the same conductor shall not be less than *six times* the trade size of the larger raceway

    Note: Six times the trade size of the largest/larger raceway is only an NEC minimum requirement, if needed the multiplier can be exceeded.

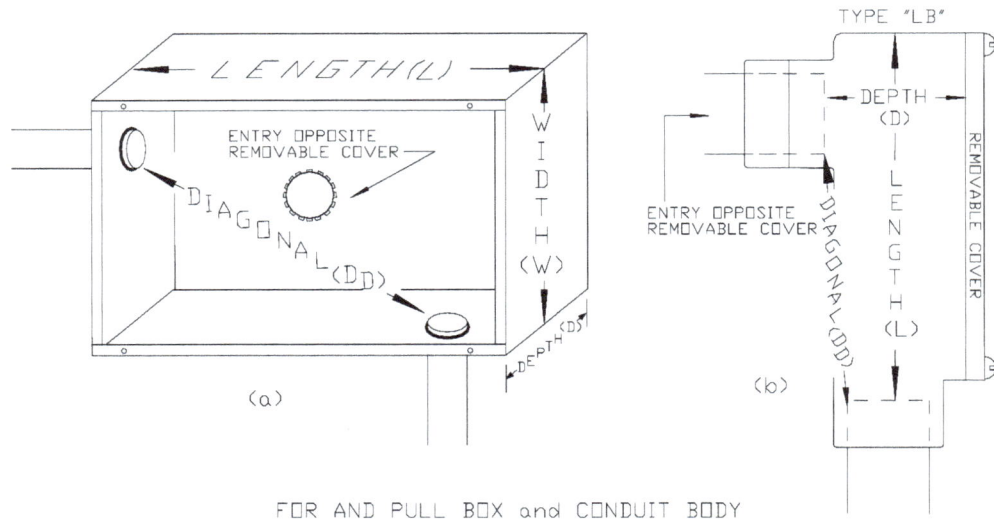

Length (L) must be 6 times the trade size of the largest raceway.
Depth (D) must be per 314.28(A)(2), Exception and Table 312.6(A).
Width (W) must be 6 times the trade size of the largest raceway (pullbox only).
Diagonal Distance (DD) must be 6 times the trade size of the largest raceway.

Figure 314.28(A)(2) - Angle Pull Box in 3-Dimensions and Applicable Conduit Body

and

(5) When the box (or conduit body) contains a removable cover opposite raceway or cable entry, the minimum depth of the box (or conduit body) is determined in accordance with Table 312.6(A). [See Figure 314.28(A)(2)]

or

refer to **Conditions 1 - 3** to determine *alternative* depth.

314.28(A)(2) - Angle Pulls or Splices (Pull and Junction Boxes and Conduit Body)
(37. - 46.) [10]

37. An angle pull box has sizes 2", 3" and 4" conduit mounted on adjacent sides of the box. Determine the minimum length and width of the box.

"Sketch the layout of the box".

Always remember, that the diagonal distance between raceway entries must not be less than six times the trade size of the largest raceway.

Since adjacent sides of the box will contain similar conduit entries, the minimum length and width of the box is determined by multiplying the size of the largest conduit being used six times *plus* adding all remaining conduit entries.

Because the largest size conduit being used is 4", the minimum length and width equals,

$$6 \times 4" + 3" + 2" = 29"$$

The minimum length and width of the box must be 29" x 29".

38. Determine the minimum overall dimensions of an angle pull box that has a 1", 2" and a 3" conduit entering one wall of the box and a 1" and 5" conduit entering an adjacent wall. The assumed conductor routing per conduit is 1" to 1" and 2" and 3" to 5".

"Sketch the layout of the box"

Unlike question No. 36. this box contains dissimilar conduit entries on adjacent walls of the box. The minimum length (L) and width (W) of the box will require being calculated separately. Therefore,

$$\mathbf{L} = 6 \times 3" + 2" + 1" = 21"$$
$$\mathbf{W} = 6 \times 5" + 1" = 31"$$

To consider the diagonal distance (DD) of the raceway enclosing the same conductors based on the largest raceway,

$$\mathbf{DD} \ (1" \ to \ 1") = 6 \times 1" \ (larger) = 6"$$
$$\mathbf{DD} \ (2" \ and \ 3" \ to \ 5") = 6 \times 5" \ (larger) = 30"$$

According to **Condition 1**

The *alternative* depth (D) is based on the largest locknut of the combination. The outer diameter of a 5" locknut is 6.75". Therefore,

$$\mathbf{D} = 4" + 6.75" = 10.75" \ (11")$$

The minimum overall dimensions are,

$$(per \ NEC) \ \mathbf{L} = 31" \ \mathbf{W} = 21" \ \mathbf{DD} \ (1") = 6" \ \mathbf{DD} \ (5") = 30"$$
$$(per \ alternative) \ \mathbf{D} = 11"$$

Use a standard size pull box based on the minimum length and width and if desired the *alternative* depth dimension.

CONSIDERING THE DIAGONAL DISTANCE (DD)

Figure 314.28(A)(2) provides visual meaning for the interpretation of the minimum required distance (diagonal) between raceway entries specified in NEC 314.28(A)(2). Usually, when the need arrives to perform such application, the cutting and spacing of holes in the walls of a desired box should be just as simple as calculating the required diagonal distance.

In the study of trigonometry there is a formula used to calculate the sides of a right triangle called the *pythagorem theorem*. The pythagorem theorem states that the square of the length of the hypotenuse (diagonal) is equal to the sum of the squares of the lengths of the sides, or in simple terms, just as any other three part formula, if the dimensions of two sides of a right triangle are known, then the third or unknown side can be determined by calculation.

The *pythagorem theorem*, as written

$$c^2 = a^2 + b^2, \text{ where } a^2 = c^2 - b^2 \text{ and } b^2 = c^2 - a^2$$

and when simplified and put in useful formula terms,

$$c = \sqrt{a^2 + b^2}, \quad a = \sqrt{c^2 - b^2} \text{ and } b = \sqrt{c^2 - a^2}$$

where **c** equals the length of the hypotenuse (diagonal distance) and **a** and **b** (derived from **c**) equals the lengths of the sides of the triangle. In formula terms, **c** is equal to the square root of sides **a** and **b** squared and sides **a** and **b** are equal to the square root of the hypotenuse **c** squared minus the square of the known side. Now, before moving on there's one important factor that always has to be realized and that is, the length of the hypotenuse **c** must always be greater than the lengths of sides **a** and **b**, otherwise the triangle will never be a closed figure.

Clearly, the adjacent corners of any square or rectangular box forms right triangles and also as a minimum at one raceway entry and a common interior wall of all conduit bodies [per Figure 314.16(C)], with the exception of Types C and E enclosures. Refer to Figure 314.28(A)(2)(a).

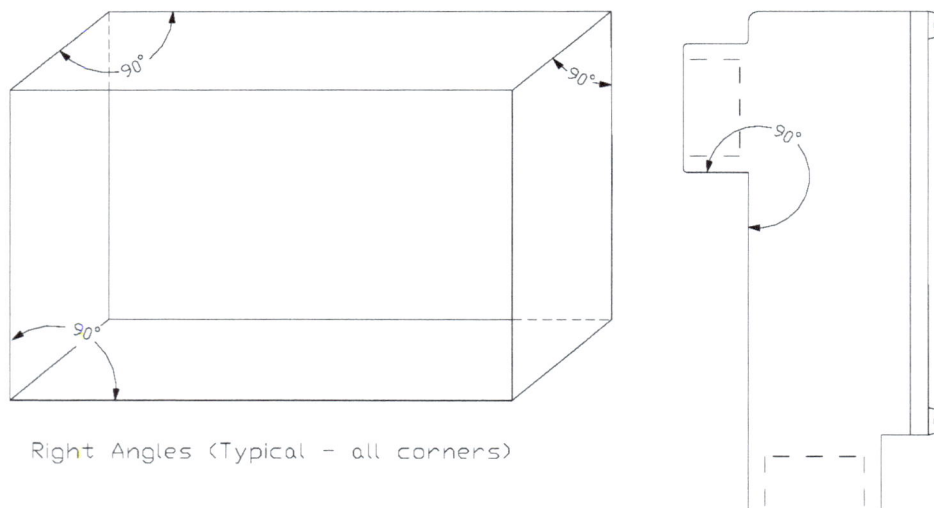

Figure 314.28(A)(2)(a)

Because of this, the pythagorem theorem can be used to ones advantage to determine dimensions or measurements that maybe otherwise difficult to apply.

Observe Figure 314.28(A)(2)(b). A 24" x 24" metal box is being used as an angle pull box for routing conductors from one 3½" raceway to another.

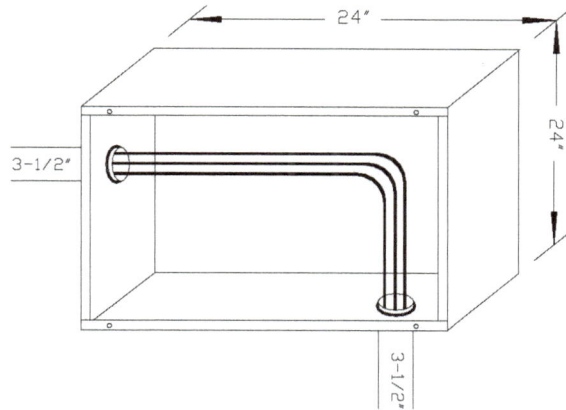

Figure 314.28(A)(2)(b)

Using the *alternative* formula in **Condition 1** as earlier covered, from the bottom edge of the 3½" locknut on the left wall of the box to the top edge of the left wall will measure 9" [4" + 5" (diameter of locknut)]. As a result, the distance from the bottom edge of the locknut to the bottom edge of the box is approximately 15". Because the diagonal distance between the two 3½" raceways is required to be a minimum of 21" (6 x 3½"), this dimension along with the distance from the bottom edge of the locknut to the bottom edge of the box can be used to determine the final straight line distance between both raceway entries by means of the pythagorem theorem. In Figure 314.28(A)(2)(c) the known dimensions are shown.

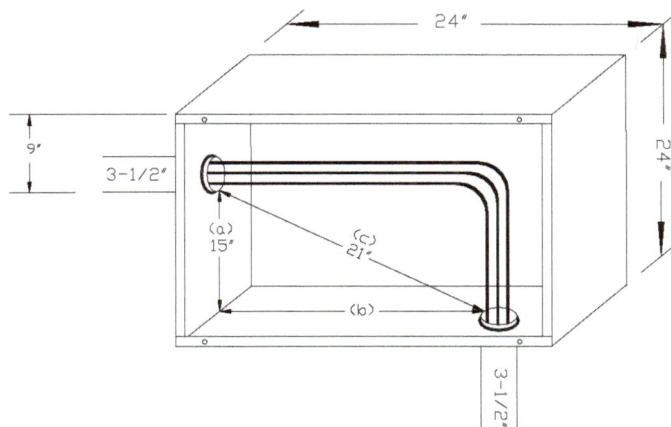

Type raceway irrelevant. Measurements taken with respect to locknuts - same results.

Figure 314.28(A)(2)(c)

The distance from the bottom edge of the left wall to the locknut of the other 3½" raceway (b) will determine how far the other raceway must be placed to maintain the *21"* diagonal distance between raceways. With the applicable formula, b = $\sqrt{c^2 - a^2}$, the unknown dimension can be determined and the spacing and cutting of the holes to install the 3½" raceways can be completed in accordance with NEC 314.28(A)(2).

$$b = \sqrt{(21")^2 - (15")^2} = 14.69" \text{ (approximately 14-11/16")}$$

To sum things up, the straight line distance from locknut edge to locknut edge of both 3½" raceways must be approximately 30", that is, one hole must be cut 15" from the bottom edge of the left wall and the other hole must be cut 14-11/16" from the left edge of the bottom wall.

The effectiveness of the pythagorem theorem is clearly obvious. Although useful, choosing a 4" measurement to derive a starting point for cutting holes in the walls of a junction or pull box is totally optional.

39. Consider the angle pull box in Figure 314.28(A)(2)-39? Determine the minimum overall dimensions of the box.

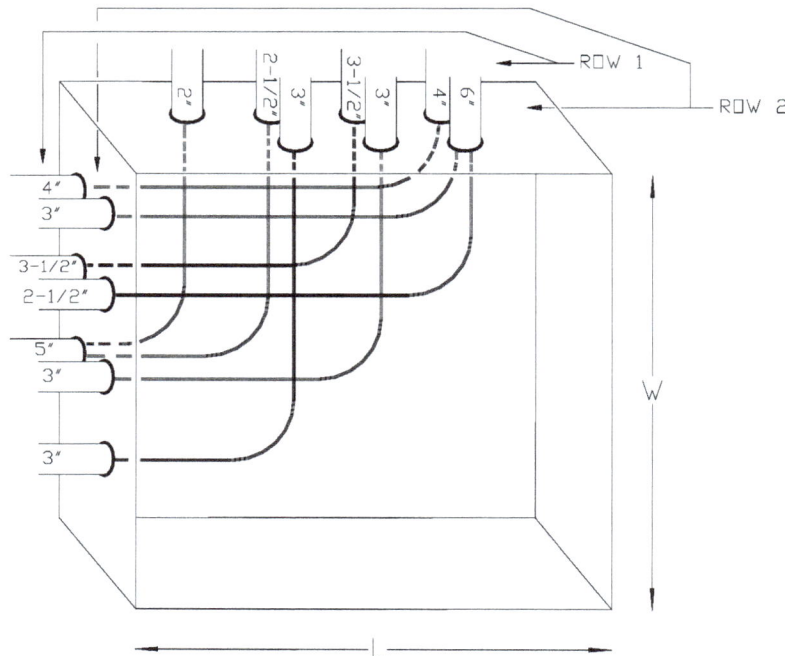

Figure 314.28(A)(2)-39

Because this box contains two rows of dissimilar sets of conduit on adjacent sides, each row will require being calculated individually. The row having the largest distance per adjacent wall will determine the length and width of the box.

Therefore,

LENGTH (**L**) - Left Wall [4" - 3½" - 5"]

(Row 1) [4" - 3½" - 5"] = 5" x 6 + 4" + 3½" = 37½"

(Row 2) [3" - 2½" - 3" - 3"] = 3" x 6 + 3" + 3" + 2½ " = 26½ "

and

WIDTH (**W**) - Top Wall

(Row 1) [4" - 3½" - 2½" - 2"] = 4" x 6 + 3½" + 2½" + 2" = 32"

(Row 2) [6" - 3" - 3"] = 6" x 6 + 3" + 3" = 42"

The minimum length of the box is determined by Row 1 (37½") and the minimum width of the box is determined by Row 2 (42").

According to **Condition 3**

The *alternative* depth (D) is determined by the wall containing the largest locknut combinations per column and adding the resulting combinations. The largest locknut combination (6", 4") originates from the top wall of the box, column 4 (from left to right). The outer diameters of a 6" and 4" locknut are 7.5938" and 5.3750", 12.9688" combined. Therefore,

D = 4" + (2" x 1) + 12.9688" = 18.97" (19")

The minimum overall dimensions are,

(per NEC) **L** = 37½" and **W** = 42" (per *alternative*) **D** = 19"

Use a standard size pull box based on the minimum length and width and if desired the *alternative* depth dimension.

40. A 24" x 24" angle pull box has 3" conduit entering the box from two locations. One location is through the wall opposite the box's removable cover and the other location is through the bottom wall adjacent to the removable cover. The conduit contains 3 - 300 kcmil THWN copper conductors. Determine the depth of the box.

"Sketch the layout of the box".

According to the *Exception* to NEC 314.28(A)(2)

When raceway enters the wall of a box opposite a removable cover the distance from the wall (depth) must comply with the column for one wire per terminal in Table 312.6(A).

Therefore,

Table 312.6(A), Column 1 - One Wire per Terminal
300 kcmil = 5" depth

Compared to **Condition 1**

The *alternative* depth (D) is determined by the wall containing the largest locknut when there is only one raceway entry. The largest locknut is 3". The outer diameter of a 3" locknut is 4.3105".

Therefore,

$$\mathbf{D} = 4" + 4.3105" = 8.3105" \ (8")$$

The minimum depth of the box must be,

(per NEC) 5" or (per *alternative*) 8"

Use a pull box that has a 5" minimum depth. An 8" depth exceeds the 5" minimum which is permitted.

41. A Type LB conduit body has a 2½" conduit entering its enclosure opposite its removable cover. The conduit contains 4 - 3/0 AWG RHW conductors. Determine the required overall dimensions for the conduit body.

Refer to Figure 314.28(A)(2). The overall dimensions of the conduit body are as calculated,

$$\mathbf{D} \ (DEPTH) = 4" \ [\text{According to Table 312.6(A) - 3/0 conductor}]$$

$$\mathbf{L} \ (LENGTH) = 6 \times 2½ \ " = 15"$$

$$\mathbf{DD} \ (DIAGIONAL \ DEPTH) = 6 \times 2½ \ " = 15"$$

The overall dimensions of the LB conduit body must meet minimum requirements where, **D** = 4", **L** = 15" and **DD** = 15".

REMINDER - Where an angle pull box or a conduit body does not provide the required diagonal distance between conduit entries, a larger enclosure must be used. Conductors that are installed in cold weather tends to stiffen and lose flexibility resulting in possible damage to the conductor's insulation when used with pull boxes or conduit bodies that only meet minimum NEC requirements. When the need to prevent insulation damage or the ease of conductor handling exist, larger enclosures are recommended.

If this material is being used for exam preparation, calculate only the dimensions required by the NEC. All *alternatives* are for desired use only.

42. A 4" raceway enters an angle pull box from the top and right walls of the box. The raceway is used to route two sets of 2/0 AWG parallel conductors. Determine the minimum dimensions of the box.

NEC 314.28(A)(2) requires the minimum distance between each raceway entry inside a box and the wall opposite each raceway entry to be no less than *six times* the trade size of the largest raceway (in a row). Because there's only one 4" raceway entry per wall, the minimum distance between the box's top wall and its opposite (bottom) wall must be (4" x 6) 24" and likewise for the box's right and opposite walls.

43. A 5" raceway enters an angle pull box from the box's left wall and a 6" raceway exits the box from the bottom wall of the box. Determine the minimum dimensions of the box.

Refer to NEC 314.28(A)(2). The minimum dimensions of the angle pull box are based on each raceway entry inside the box and the wall opposite the entry. Because there are two different size raceways entering the box from adjacent walls, the minimum dimensions are as follows:

Distance from left to opposite wall (5" raceway) must be (5" x 6) 30".
Distance from bottom to opposite wall (6" raceway) must be (6" x 6) 36".

44. Two adjacent sides of an angle pull box contain two rows of raceway entries. The raceway entries are as follows:

SIDE "A"	SIDE "B"
Row 1 - 3", 3", 2", 2", 4"	Row 1 - 3½", 2½", 1½", 1¼"
Row 2 - 3½", 2½", 1½", 1¼"	Row 2 - 6", 5", 4"

Conductors no smaller than 4 AWG will be routed through designated raceway entries. Determine the minimum size angle pull box required for this installation.

In addition to requiring the distance between a raceway entry and an opposite wall to be *six times* the trade size of the largest raceway entry, NEC 314.28(A)(2) also requires the distance to be extended by adding the diameters of any additional raceway entries within a row that's on the same wall. Where more than one row will exist on the wall of an angle pull box, NEC 314.28(A)(2) further requires each row to be calculated individually and the single row providing the maximum distance to be used as the required dimension for that particular wall.

Evaluating each row of both sides,

SIDE A

Row 1 - 4" x 6 + 3" + 3" + 2" + 2" = 34" - (maximum distance)
Row 2 - 3½" x 6 + 2½" + 1½" + 1¼" = 26.25"

SIDE B

Row 1 - 3½" x 6 + 2½" + 1½" + 1¼" = 26.25"
Row 2 - 6" x 6 + 5" + 4" = 45" - (maximum distance)

Based on the calculated results, the angle pull box cannot be sized *no less* than 45" x 34" for this installation.

45. Determine the required distance between the raceway entries in question No. 43.

According to NEC 314.28(A)(2), the distance (diagonal) between raceway entries enclosing the same conductor shall not be less than *six times* the trade size of the larger raceway. With this as a requirement, the (diagonal) distance between the raceway entries must be 36" (6" x 6).

46. Refer to question No. 44. If the 5" raceway in row 2 of side B is used to enclose two sets of parallel conductors originating from the two 2" raceways in row 1 of side A, how much distance is required to be maintained between the raceways?

Again, the distance (diagonal) between raceway entries enclosing the same conductors must not be less than *six times* the trade size of the larger raceway. As a result, the (diagonal) distance between the raceway entries must be 30" (5" x 6).

314.28(A)(2), *Exception* - Angle Pulls (Pull and Junction Boxes and Conduit Bodies)
               (47. - 48.) [2]

47. A Type LB conduit body is being used to route 3 - 2 AWG conductors. A 2" raceway is attached to both hubs of the enclosure. What are the minimum dimensions required for this enclosure?

According to the requirements of NEC 314.28(A)(2), the distance between the raceway entry opposite the wall of the conduit body and the distance between each raceway entry must be 12" (6 x 2"). As for the raceway entry opposite the enclosure's removable cover, the distance is determined per Table 312.6(A), for one wire per terminal. Based on Table 312.6(A) the distance from the raceway entry to the removable cover must be 2½".

48. A 3" conduit containing 4 - 400 kcmil conductors enters a junction box opposite the box's removable cover. Another 3" conduit entering the bottom wall of the box is used to complete the routing of the conductors. Determine the minimum dimensions of the box.

According to the *Exception* to 314.28(A)(2), when raceway enters the wall of a box opposite a removable cover the distance from the wall must comply with the column for one wire per terminal in Table 312.6(A). Referring to Column 1 (one wire per terminal) of Table 312.6(A), the distance between the wall opposite the removable cover must be 6" based on the use of the 400 kcmil conductors.

The distance between the 3" raceway entering the bottom wall of the box to the opposite wall must be 18" (3" x 6) as well as the distance (diagonal) between the raceway entries.

THE U-PULL

- ### As defined

    A box used where raceway(s) of similar or dissimilar sizes are mounted on the same wall of the box running parallel to each other to reverse the direction of the enclosed conductors.

- ### As sized

    When sizing a box to conduct a U-pull [See Figure 314.28(A)(2)(d)] NEC 314.28(A)(2) requires,

    (1) The distance (width) between each point where raceway enters the inside of the box and the opposite wall of the box must not be less than *six times* the trade size (metric designator) of the largest raceway and the distance (width) shall be increased for additional entries by the amount of the sum of the diameters of all other raceway entries on the same wall of the box, where

    > **W** = 6 x trade size (metric designator) of largest raceway *plus* diameters of other raceway

    (2) When there are multiple rows on the same wall of the box, each row much be calculated individually *and* the single row that provides the maximum distance shall be used.

    (3) The linear distance (**LD**) between raceway entries enclosing the same conductor must not be less than *six times* the trade size (metric designator) of the largest raceway.

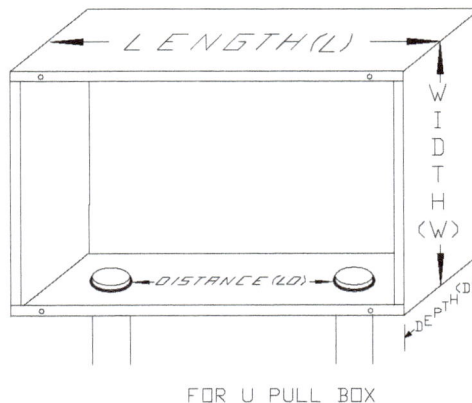

FOR U PULL BOX

Linear Distance (LD) must be 6 times the diameter of the largest raceway measured from entry to entry.

Figure 314.28(A)(2)(d) - U-Pull Box in 3-Dimensions

Since the NEC does not provide a minimum length and depth for a U-Pull consider the following *alternatives* (non-NEC directives).

ALTERNATIVES

(1) Because the length (L) of the box must have enough space to satisfy step (3), as a rule of thumb, add the distance found in step (3) *plus* the diameter of the largest locknut *twice* and allow at least a 6" space between the largest locknuts and the outer walls parallel to the raceway, where,

$$L = 6 \text{ x trade size of largest raceway} + (2 \text{ x diameter of largest locknut}) + 12"$$

(2) The depth (D) of the box can be determined by **Conditions 1**, **2**, or **3**.

314.28(A)(2) - U Pulls (Pull and Junction Boxes and Conduit Bodies) (49. - 51.) [3]

49. A U-pull box has 3" and 3½" conduit mounted on the same wall of the box. Determine the minimum distance between the conduit and the minimum width of the box.

"Sketch the layout of the box".

Since the box contains different conduit entries, the minimum distance (linear) between the conduit and the width of the box are both determined by multiplying the size of the largest conduit being used by six (6). The width of the box however, must also include all remaining conduit entries.

Per larger size conduit, the minimum distance between conduit must be, 6 x 3½ " = 21" and the minimum width of the box must be, 21" + 3" = 24".

The minimum distance (between conduits) and width of the box must be 21" and 24", respectively.

50. What are the minimum overall dimensions of a U-pull box that has 3½" and 4" conduit mounted parallel on the right wall of the box?

"Sketch the layout of the box".

The minimum distance and width can be determined using the same methods as those used in question No. 49.

Per larger size conduit, the minimum distance (LD) between conduit must be, 6 x 4" = 24" and the minimum width of the box must be, 24" + 3½" = 27½".

The *alternative* length (L) can be determined based on the method and formula previously given in step (1) (**ALTERNATIVES**). The largest locknut is 4". The outer diameter of a 4" locknut is

5.3750". Therefore,

$$L = 6 \times 4" + (2 \times 5.3750") + 12" = 46.75" \,(47")$$

According to **Condition 1**

The *alternative* depth (D), is based on the largest locknut of the combination which again is 4". Therefore,

$$D = 4" + 5.3750" = 9.375" \,(9")$$

The minimum overall dimensions are,

(per NEC) **LD** = 24" **W** = 27½" (per *alternative*) L = 47" **D** = 9"

Use a standard size pull box based on the minimum overall dimensions above.

51. A junction box is being used to make a U-pull consisting of 6 - 250 kcmil conductors. The bottom wall of the box contains two parallel 4" raceways. Determine the minimum size junction box needed to make the U-pull.

The requirements for determining the size of the junction box being used for making U-pulls are the same as those for angle pulls according to NEC 314.28(A)(2). The minimum distance between each raceway entry inside a box and the wall opposite each raceway entry must not be less than *six times* the trade size of the largest raceway (in a row).

$$4" \times 6 = 24"$$

The distance between the bottom wall where the 4" raceways are mounted and the opposite wall must be no less than 24".

REMINDER - Where an angle pull box or a conduit body does not provide the required diagonal distance between conduit entries, a larger enclosure must be used. Conductors that are installed in cold weather tends to stiffen and lose flexibility resulting in possible damage to the conductor's insulation when used with pull boxes or conduit bodies that only meet minimum NEC requirements. When the need to prevent insulation damage or the ease of conductor handling exist, larger enclosures are recommended.

If this material is being used for exam preparation, calculate only the dimensions required by the NEC. All *alternatives* are for desired use only.

314.28(A)(2) - Angle Pull Box (Pull and Junction Boxes and Conduit Bodies)
(when cables are installed)

52. A 2-3 nonmetallic sheathed cable with an 8 AWG bare equipment grounding conductor will require being spliced in a junction box. One part of the cable will enter the box from the top wall while the other part of the cable will enter the box from the left wall. The 2 AWG conductors of

the 2-3 cable have type TW insulation. Determine the minimum size junction box needed.

"Sketch the layout of the box"

NEC 314.28(A)(2) states, when transposing cable size into raceway size in NEC 314.28(A)(1) and (A)(2), the minimum trade size raceway required for the number and size of conductors in the cable shall be used.

Therefore, the first thing that's required to size the angle pull (splice) box is to determine the minimum size raceway needed. The minimum size raceway is then determined by calculating the cross-sectional area of each conductor enclosed in the cable.

Referring to **Table 5** of **Chapter 9**, the cross-sectional area of a 2 AWG TW conductor is .1333 in^2. Referring to **Table 8** of **Chapter 9**, the cross-sectional area of a bare 8 AWG conductor is .013 in^2. The total cross-sectional area of all conductors is now determined.

$$2 \text{ AWG} - 3 \times .1333 \text{ in}^2 = .3999 \text{ in}^2$$
$$(\text{solid}) \; 8 \text{ AWG} - 1 \times .0130 \text{ in}^2 = \underline{.0130 \text{ in}^2}$$
$$.4129 \text{ in}^2 \; (\text{total})$$

According to the **Notes to Tables** of **Chapter 9**, **Note 9**; a multiconductor cable of two or more conductors shall be treated as a single conductor for calculating percentage conduit fill area. Of all raceways listed in **Table 4** of **Chapter 9**, Type EB and Schedule 80, Rigid PVC conduits are the only raceways requiring a trade size larger than 1" to enclose the non-metallic cable at a 53 percent conduit fill area. Therefore, based on the smallest (minimum) size raceway (1") that can be used to enclose the conductors the distance between the cable entries and the walls opposite the entries are required to be six times the 1" trade size of the smallest size raceway (6 x 1"). As a minimum, a 6" x 6" junction box is needed.

For applications pertaining to multiconductor cable conduit fills refer to Chapter 9 (Volume 3).

314.28(A)(1) and (2) (Pull and Junction Boxes and Conduit Bodies) - Combination Pulls

53. A junction box is being used to make a combination of pulls. A 4" raceway is mounted on the right and left walls of the box to conduct a straight pull for 8 - 3/0 AWG conductors. Two 3½" raceways emerge from the top and right walls of the box containing 5 - 250 kcmil conductors. The bottom wall of the box contains a 2½" raceway along with two 3" raceways that are used to route 7 - 4/0 AWG conductors between the two raceways. A 2½" raceway containing 3 - 300 kcmil conductors enters the wall opposite the box's cover and couples with the other 2½" raceway to complete the routing of the conductors. Taking all combinations of pulls into consideration, what size junction box is needed for this installation?

"Sketch the layout of the box".

To begin, let's consider each pull individually, focusing on all raceway entries per NEC 314.28(A)(2) to adequately size the needed junction box.

Straight pull - 4" x 8 = **32"** [minimum distance (length) from *left* to *right* walls]
Angle pull - 4" x 6 + 3½" = 27½" [minimum distance from *right* to *left* walls]
Angle/U-pull - 3" x 6 + 3" + 2½" = 23½" [minimum distance from *bottom* to *top* walls]
3½" x 6 = 21" [minimum distance from *top* to *bottom* walls]
Wall to cover - 5" [per Table 312.6(A) based on 300 kcmil conductors]

The junction box is sized based on the greater calculated distances between the left and right walls and the top and bottom walls. As a minimum, the junction box can be no less than 32" x 23½" and have a wall to cover distance of 5."

314.28(A)(3) - Smaller Dimensions (Pull and Junction Boxes and Conduit Bodies)

54. A 2" metal LB conduit body is marked "max wire size 3-4/0 AWG" while appropriately attached to a 2½" metal conduit enclosing the same combination of conductors. Can the conduit body be used with the oversized conduit for making an angle pull?

If the conduit body meets all of the requirements of NEC 314.28(A)(2) it can be used to make an angle pull. However, according to NEC 314.28(A)(3) the smaller size conduit body can be used with a conduit that has a capacity to enclose more conductors than the conduit body providing the conduit body is listed for and is permanently marked with the maximum number (**3**) and the maximum size (**4/0 AWG**) of the conductors permitted.

Based on the applicable tables of **Informative Annex C** that list metal conduit, all 2½" metal raceways has the capacity to enclosed more than three 4/0 AWG conductors regardless of insulation type.

314.71(A) - For Straight Pulls (Size of Pull and Junction Boxes, Conduit Bodies and Hand hole Enclosures [Systems Over 600 volts])

55. A medium voltage cable containing shielded conductors enters both ends of a straight pull box. The diameter of the cable is 1.25". Determine the minimum length of the straight pull box.

The minimum length of the box must be *48 times* the diameter of the cable entering the box when the cable will contain shielded conductors. Therefore, the minimum length of the box is required to be 60" (48" x 1.25"). If the cable was unshielded the minimum length of the box would be required to be *32 times* the diameter of the cable.

314.71(B)(1) - Distance to Opposite Wall (For Angle or Pulls [Systems Over 600 volts])

56. An angle pull box is used to route three sets of non-shielded/non-lead covered medium voltage cable. The cables will enter the box through the bottom wall and extend through the

right interior wall of the box. The outer diameters of the cables are .775", 1.086" and 1.337". Determine the minimum dimensions of the box.

Because the cables are non-shielded/non-lead covered the distance from the cable entries to the opposite wall of the box is permitted to be reduced to *24 times* the outside diameter of the largest cable according to NEC 314.71(B)(1), *Exception No. 2* opposed to 36 times per NEC 314.71(B)(1).

The minimum distance between both walls of the cable entries must be 24 times the outside diameter of the largest cable plus the outside diameters of all remaining cables entering the same wall of the box. Therefore,

$$24 \times 1.337" + 1.086" + .775" = 33.95" \text{ (rounded-off} - 34")$$

The distances between the cable entries and the walls opposite the entries are required to be as a minimum, 34" (34" x 34").

314.71(B)(1), *Exception No. 1* - For Angle Pulls (Systems Over 600 volts)

57. If the cable in question No. 55. was routed through a box opposite the box's removable cover, what would be the required minimum distance between the box and cover?

Per NEC 314.71(B)(1) - *Exception No. 1*, for such installation the bending radius for the conductors shall be permitted to be based on the provisions of NEC 300.34. For the (multiconductor) cable, the bending radius must be *7 times* the overall diameter of the cable being that the diameter of the individually enclosed shielded conductors is unknown. Therefore,

$$1.25" \times 7 = 8.75"$$

314.71(B)(2) - Distance Between Entry and Exist (For Angle or U Pulls [Systems Over 600 volts])

58. Determine the minimum distance between each cable entry/exit in question No. 56.

According to NEC 314.71(B)(2), the distance between a cable entry and its exits from the box shall not be less than *36 times* the outside diameter, over sheath, of that cable. Refer to the following:

Diameter(s)		Minimum distance *(diagonal)*
.775" x 36	=	27.9"
1.086" x 36	=	39.1"
1.337" x 36	=	48.1"

ARTICLE 326 - Integrated Gas Spacer Cable: Type IGS

326.116 - Conduit (1.- 2.) [2]

1. An integrated gas spacer cable contains 20 solid aluminum rods at ½" in diameter. If the overall cross-sectional area of the cable is 40 percent greater than the total cross-sectional area of the rods (conductors), what size polyethylene conduit is required to enclose the conductors?

$$\text{Diameter of single rod} = \text{½" (.5in.)}$$

Using the geometrical formula for determining the area of a circle where, $A = \pi \times (d/2)^2$, the cross-sectional area of the 20 rods is,

$$A = \pi \times (.5 \text{ in.}/2)^2 \times 20 = 3.925 \text{ in}^2 \text{ (where } \pi = 3.14)$$

Based on the overall cross-sectional area of the cable being 40 percent (.40) greater than the total cross-sectional area of the rods,

$$3.925 \text{ in}^2 \times 1.40 = 5.495 \text{ in}^2$$

Converting the original (area of a circle) formula to derive the diameter of the desired polyethylene conduit where, $d = \sqrt{(A/\pi)} \times 2$, the conduit's internal diameter must be as a minimum, 2.65 in. Observe,

$$d = \sqrt{(5.495 \text{ in}^2/3.14)} \times 2$$
$$= 2.65 \text{ in.}$$

Based on the calculated results, per Table 326.116, a 3" polyethylene conduit which has a 2.886 in. actual inside diameter must be used.

2. Based upon the calculated cross-sectional area of the cable at 1.40 percent, if the cable was installed in Schedule 40 PVC, what size would be required?

According to the **Notes to Tables** of Chapter 9, Note 9; multiconductor cable must be treated as a single conductor for calculating conduit fill area. Referring to Table 4 of Chapter 9, based on the calculated cross-sectional area of the cable at 5.495 in² and applying the 53 percent fill capacity, 4" Schedule 40 PVC is needed which has a cross-sectional area of 6.654 in².

ARTICLE 330 - Metal-Clad Cable: Type MC

330.24(A)(2)- Smooth Sheath (Bending Radius)

1. A metal-clad cable secured to the side of a wooden stud is required to be routed through the stud prior to terminating in a junction box. If the diameter of the cable is 1.125" and the cable has a smooth metallic outer sheath, determine the cable's minimum bending radius.

Referring to NEC 330.24(A)(2), the minimum bending radius must be 12 times the external diameter of the cable where the diameter is more than ¾" but not more than 1½". As a result,

$$12 \times 1.125" = 13.5"$$

The minimum bending radius must be 13.5".

330.24(B) - Interlocked-Type or Corrugated Sheath (Bending Radius)

2. If the outer diameter of metal-clad cable with interlocked-type armor is .625", determine the bending radius of the cable.

According to NEC 330.24(B), the bending radius of metal-clad cable with interlocked-type armor must be seven times the external diameter of the cable. As calculated,

$$7 \times .625" = 4.375"$$

The bending radius of the cable must be at least 4.375".

330.24(C) - Shielded Conductors

3. Refer to question No. 1. of Article 300 (300.34 - Conductor Bending Radius).

The requirements and application are the same for multiconductor cables and individually shield conductors rated for over 600 volts.

ARTICLE 334 - Nonmetallic-Sheathed Cable: Types NM (Romex), NMC, and NMS

334.24 - Bending Radius

1. The diameter of a 12-2 nonmetallic-sheathed cable is 3/16". Determine the bending radius of the cable.

According to NEC 334.24, the bending radius of the cable must be no less than 5 times the diameter of the cable. After converting the fractional measurement 3 /16" into a decimal measurement of .1875" the bending radius can be determine. Therefore,

$$.1875" \text{ x } 5 = .9375"$$

where .9375" is the same as 15/16". As a result the bending radius cannot be less than 15/16".

ARTICLE 348 - Flexible Metal Conduit: Type FMC

Table 348.22 - Maximum Number of Insulated Conductors in Flexible Metal Conduit (1.- 2.)[2]

1. How many 14 AWG XHHW conductors can be installed in 3/8" flexible metal conduit (FMC) containing outer fittings?

According to Table 348.22, three (3) 14 AWG XHHW conductors can be enclosed in a 3/8" FMC. See NEC 350.22(B) for Liquidtight Flexible Metal Conduit: Type LFMC.

2. How many 14 AWG equipment grounding conductors can be installed with the conductors in question No. 1?

Referring to the footnote to Table 348.22, one insulated, covered, or bare equipment grounding conductor is permitted to be installed.

ARTICLE 352 - Rigid Polyvinyl Chloride Conduit: Type PVC

Table 352.44 - Expansion Characteristics of PVC Rigid Nonmetallic Conduit......) (1.- 2.)[2]

1. When 100' of Type PVC is installed in an area where the temperature is expected to range on the average between 75°F and 125°F, rigid nonmetallic conduit will expand how many inches?

According to Table 352.44, when there is a temperature change of 50°F (75°F - 125°F) which would be worst case for this installation, 100' of Type PVC is expected to expand 2.03".

2. If the Type PVC per question No. 1. is ran 255' in the same environment, how many inches will the conduit expand?

Since the conduit will expand 2.03" per 100' when there is a temperature change of 50°F, at 255' the distance is 2.55 times greater. Therefore, the answer is,

$$2.55 \times 2.03" = 5.1765"$$

which was derived by the use of proportions. Observe,

$$\frac{2.03"}{100'} = \frac{x}{255'} \quad \text{where, } x = \frac{255' \times 2.03"}{100'} = 5.1765"$$

ARTICLE 354 - Nonmetallic Underground Conduit with Conductors: Type NUCC

Table 354.24 - Minimum Bending Radius for Nonmetallic Underground Conduit with Conductors (NUCC)

1. Determine the minimum bending radius of Type NUCC for trade sizes ¾", 1¼", 2½" and 4".

According to NEC 354.24, the radius of the curve of the centerline of conduit bends involving Type NUCC shall not be less than shown in Table 354.24.

Per Table 354.24, trade sizes ¾", 1¼", 2½" and 4" requires a minimum bending radius of 12", 18", 36" and 60" respectively.

ARTICLE 355 - Reinforced Thermosetting Resin Conduit: Type RTRC

Table 355.44 - Expansion of Characteristics of Reinforced Thermosetting Resin Conduit (RTRC)

Same application as Table 352.44. See question Nos. 1. and 2.

AUXILIARY GUTTERS AND WIREWAYS

Auxiliary gutters are mostly used for making short connections to disconnect switches, panel boards, and meter cans and are only permitted to enclose conductors or busbars. Switches, overcurrent devices and other similar type equipment are not permitted to be enclosed in auxiliary gutters. Unlike wireways, auxiliary gutters are not permitted to extend more than 30 feet beyond the equipment they supplement.

An auxiliary gutter carries the same 20 percent allowance as that of a wireway in terms of fill capacity. However, as for the number of current-carrying conductors and the allowed cross-sectional area where splice and tap conductors are included, the requirements differ based on the use of metal and non-metal enclosures.

When the need exist to run large conductors or a large number of conductors to a distance location or to several locations, it is quite common to use a wireway in place of raceway. A wireway is permitted to be installed either indoors or outdoors where listed for such purpose and may be extended to any distance.

Wireways are commonly manufactured in standard lengths of 1 to 5 feet and 10 feet and from 2½ inches x 2½ inches up to 12 inches x 12 inches in cross-sectional area. Accessories such as fittings, elbows, couplings, crossovers, and end sections are made available to assemble a wireway structure.

Although auxiliary gutters and wireways are very similar in appearance, an auxiliary gutter function similar to that of a junction box while a wireway function very similar to that of surface raceway.

ACCORDING TO THE NEC

SIZING AUXILIARY GUTTERS AND WIREWAYS

The following NEC sections are applicable for understanding how auxiliary gutters (Article 366) and wireways (Articles 376 and 378) are sized and selected.

As applied to Auxiliary Gutters (Article 366)

NEC 366.12 - Use Not Permitted (Auxiliary Gutters) - An auxiliary gutter shall not extend no more than *30 ft.* beyond the equipment it supplements.

Exception - Auxiliary gutters used for elevators are not subject to the 30 feet limitation. Refer to NEC 620.35.

NEC 366.22(A) - Sheet Metal Auxiliary Gutters (Number of Conductors) - The sum of the cross-sectional areas of all contained conductors at *any* cross section of a sheet metal auxiliary gutter shall not exceed *20 percent* (.20) of the interior cross-sectional area of the gutter, see Figure 366.22(A)(B)*. Conductors that are used for signal circuits, or controller circuits (for starting duty only) between motors and their starters are not considered current-carrying, yet are included within the *20 percent* (.20) interior cross-sectional area limitation. The adjustment factors as outlined in NEC 310.15(B)(3)(a) are only applicable when the number of current-carrying conductors including neutral conductors classified as current-carrying per NEC 310.15(B)(5) exceeds *30*.

NEC 366.22(B) - Nonmetallic Auxiliary Gutters (Number of Conductors) - The sum of the cross-sectional areas of all contained conductors at *any* cross section of the nonmetallic auxiliary gutter shall not exceed *20 percent* (.20) of the interior cross-sectional area of the gutter, see Figure 366.22(A)(B).

*Figure 366.22(A)(B) only serves as a means to illustrate a geometrical (rectangular object) fill at 20 percent opposed to a 20 percent fill capacity pertaining to conductors.

GUTTERS and WIREWAYS

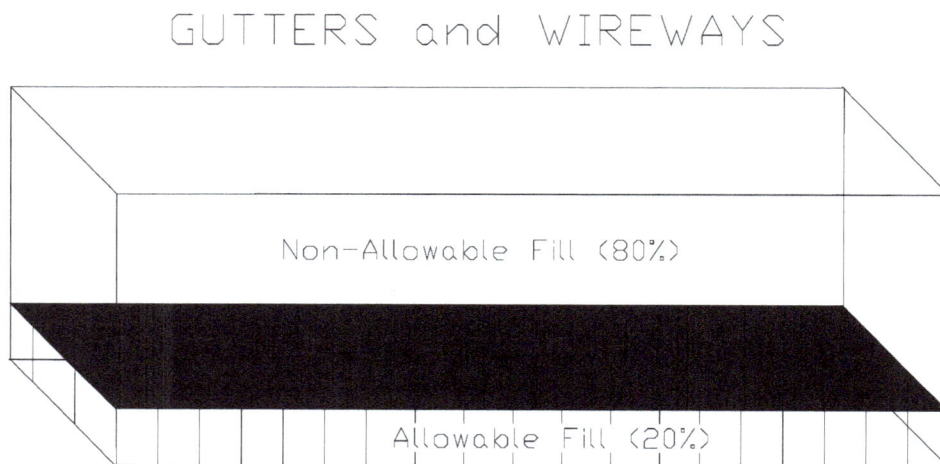

Non-Allowable Fill (80%)

Allowable Fill (20%)

Only 20% of the cross-sectional area of auxiliary gutters and wireways are allowed to be filled with conductors.

Figure 366.22(A)(B) - Conductor's allowable fill of gutters and wireways

NEC 366.23(A) - Sheet Metal Auxiliary Gutters (Ampacity of Conductors) - NEC 310.15(B)(3)(a) does not apply when the number of current-carrying conductors contained in a sheet metal auxiliary gutter is less than *30*. When bare copper bars are used in sheet metal auxiliary gutters, the ampacity of the copper bar is limited to *1000 amperes* per square inch (1000A/in^2) of the cross section of the conductor (bar). For aluminum bars, the ampacity is limited to *700 amperes* per square inch (700A/in^2) of the cross section of the conductor (bar).

NEC 366.23(B) - Nonmetallic Auxiliary Gutters (Ampacity of Conductors) - The adjustment factors as outlined in NEC 310.15(B)(3)(a) are applicable when the number of current-carrying conductors including neutral conductors classified as current-carrying per NEC 310.15(B)(5) exceeds *three* (3).

NEC 366.30(A) - Sheet Metal Auxiliary Gutters (Securing and Support) - Must be supported and secured throughout their entire length at intervals not exceeding 5 ft.

NEC 366.30(B) - Nonmetallic Auxiliary Gutters (Securing and Support) - Must be supported and secured at intervals not exceeding 3 ft. and at each end or joint, unless listed (approved) for other support intervals. In no case shall the distance between supports exceed 10 ft.

NEC 366.56(A) - Within Gutters (Splices and Taps) - Splices or taps are permitted if they are accessible. Conductors, including splices and taps must not fill (occupy) more than *75 percent* (.75) of the gutter's interior cross-sectional area. See Figure 366.56(A)*.

*Figure 366.56(A) only serves as a means to illustrate a geometrical (rectangular object) fill at 75 percent opposed to a 75 percent fill capacity pertaining to conductors.

AUXILIARY GUTTERS

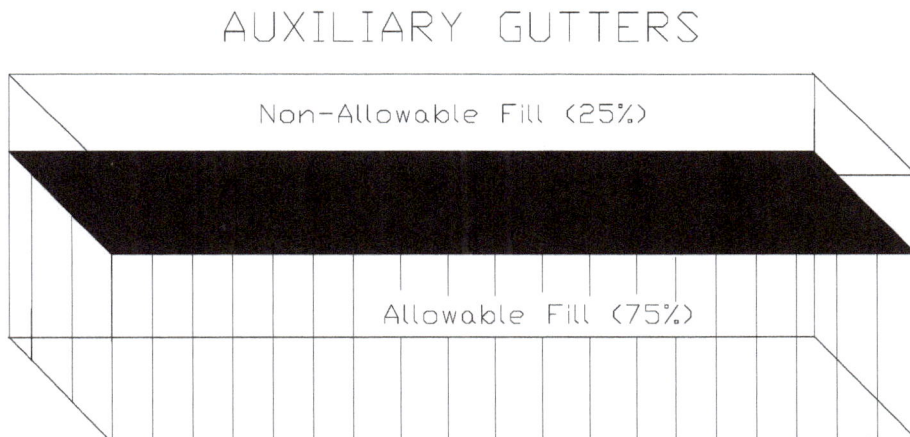

Non-Allowable Fill (25%)

Allowable Fill (75%)

The allowable conductor fill in auxiliary gutters is permitted to occupy up to 75% of the gutter's cross-sectional area when splices and taps are included with other conductors.

Figure 366.56(A)

NEC 366.56(B) - Bare Conductors (Splices and Taps) - When taps are made from bare conductors, such as busbars, the tap must leave the gutter opposite their point of terminal connection to the busbar. Taps must not come into contact with any bare current-carrying parts of different potential. See Figure 366.56(B).

AUXILIARY GUTTER

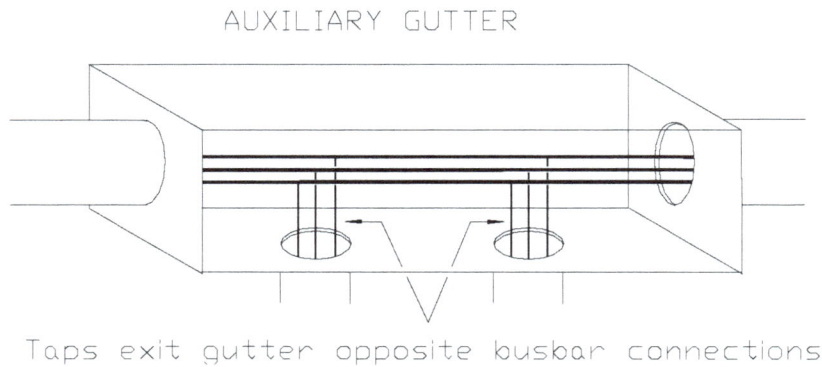

Taps exit gutter opposite busbar connections

Figure 366.56(B) - Tap conductors leaving gutter opposite point of connection

NEC 366.56(D) - Overcurrent Protection (Splices and Taps) - Tap conductors within an auxiliary gutter are subject to the tap rules of NEC 240.21.

NEC 366.58(A) - Deflected Insulated Conductors (Insulated Conductors) - When conductors are deflected (the required bending of the conductor*) in an auxiliary gutter whether at the point of entry or exit, or when the direction of the deflection exceeds *30 degrees* (30°), the width of the gutter is determined per Table (NEC) 312.6(A). In addition to NEC 366.22(A), (B) and 366.56(A), NEC 366.58(A) also requires auxiliary gutters to be sized according to the deflection of insulated conductors. See Figure 366.58(A).

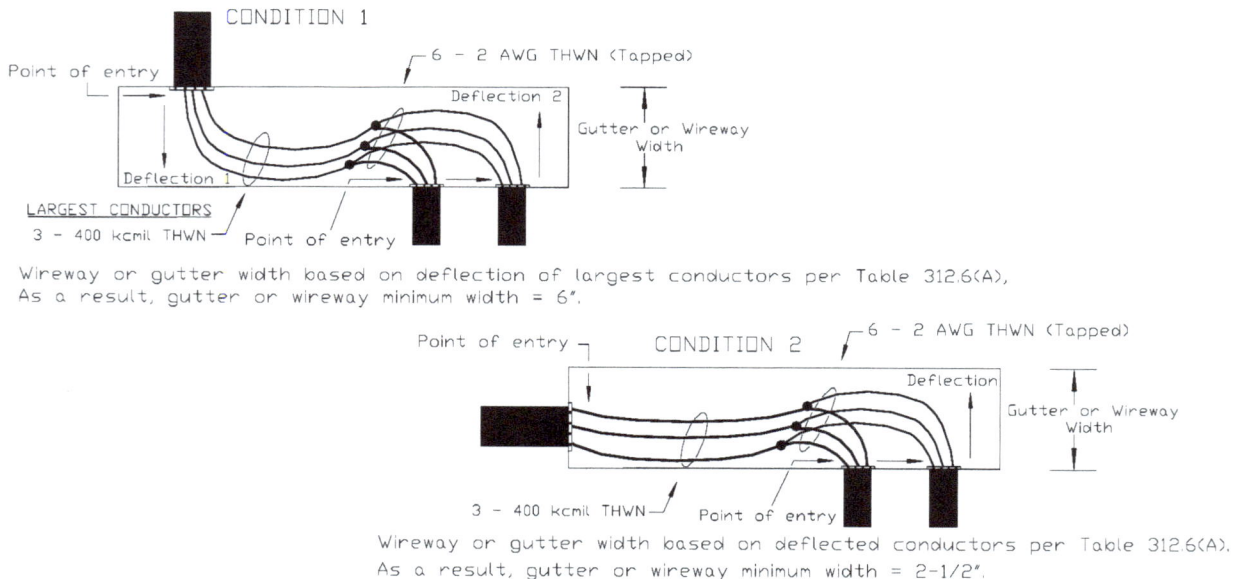

Figure 366.58(A)

*The required bending of a conductor is a reflection of the conductor's bending radius from the point of entry to the point of termination, splice, tap, or exit. Such requirement protects the insulation of a conductor from failure due to over extending on the outside of the bend, or from crimping on the inside of the bend.

NEC 366.58(B) - Auxiliary Gutters Used as Pull Boxes (Insulated Conductors) - Where insulated conductors 4 AWG or larger are *pulled through* an auxiliary gutter, the distance between raceway and cable entries enclosing the same conductor shall not be less than required in NEC 314.28(A)(1) for straight pulls and NEC 314.28(A)(2) for angle pulls. See Figure 366.58(B).

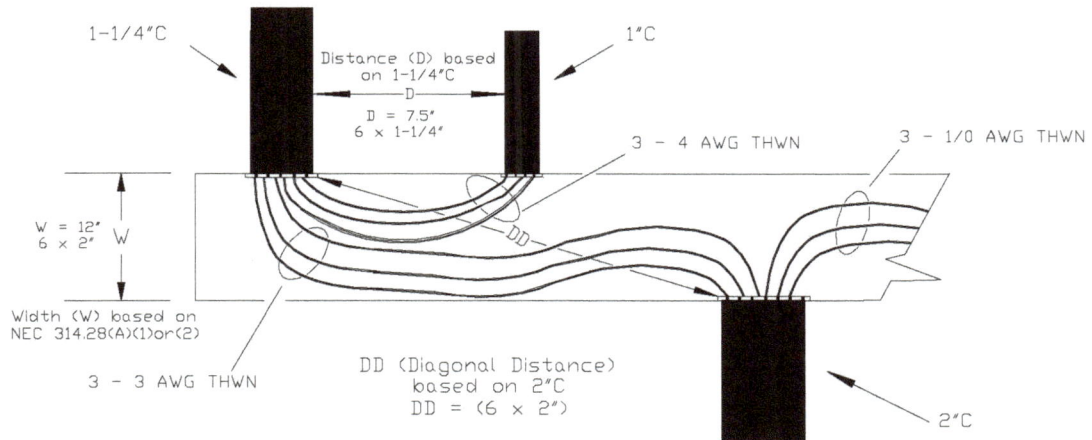

Figure 366.58(B) - Minimum distance between gutter (or wireway) entries

NEC 366.60 - Grounding (Insulated Conductors) - Metal auxiliary gutters shall be connected to an equipment grounding conductor(s), to an equipment bonding jumper, or to the grounded conductor where permitted or required by NEC 250.92(B)(1) or 250.142.

NEC 366.100(E) - Clearance of Bare Live Parts (Construction) - Bare conductors must be securely and rigidly fastened in place. The minimum clearance between bare current-carrying metal parts, such as busbars must not be less than 2" apart at fastened points and not less than 1" apart between the bare current-carrying metal parts and any metal surface. Adequate allowance must be made for the expansion and contraction of busbars. See Figure 366.100(E).

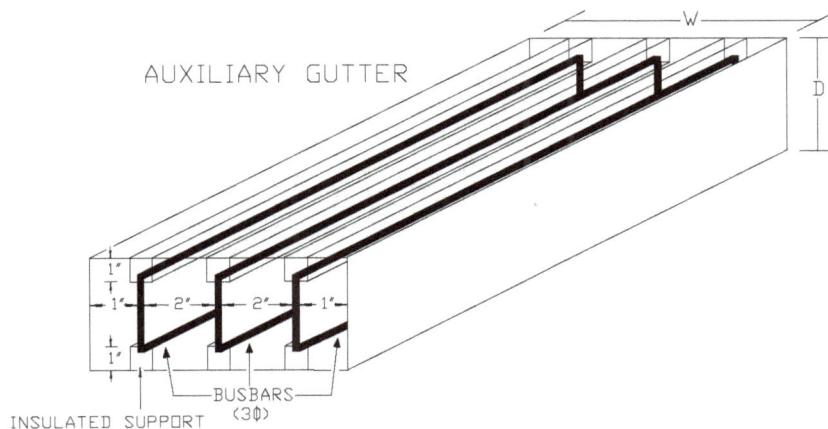

Figure 366.100(E) - Clearance of Bare Live Parts

As applied to Metal Wireways (Article 376) and Nonmetallic Wireways (Article 378)

NEC 376.21 [NEC 378.21] - Size of Conductors - The design of the (metal/nonmetallic) wireway shall determine the maximum size conductors permitted.

NEC 376.22(A) - Cross-Sectional Areas of Wireway (Number of Conductors and Ampacity) - The sum of the cross-sectional areas of all contained conductors at *any* cross section of a metal wireway shall not exceed *20 percent* (.20) of the interior cross-sectional area of the wireway, see Figure 366.22(A)(B).

NEC 376.22(B) - Adjustment Factors (Number of Conductors and Ampacity) - The adjustment factors as outlined in NEC 310.15(B)(3)(a) are only applicable when the number of current-carrying conductors including neutral conductors classified as current-carrying per NEC 310.15(B)(5) exceeds *30*. Conductors that are used for signal circuits, or controller circuits (for starting duty only) between motors and their starters are not considered current-carrying.

NEC 378.22 - Number of Conductors - The sum of the cross-sectional areas of all contained conductors at *any* cross section of the nonmetallic wireway shall not exceed *20 percent* (.20) of the interior cross-sectional area of the wireway. Conductors that are used for signal circuits, or controller circuits (for starting duty only) between motors and their starters are not considered current-carrying. The adjustment factors as outlined in NEC 310.15(B)(3)(a) are applicable for all current-carrying conductors that are included within the *20 percent* (.20) cross-sectional area limitation. See Figure 366.22(A)(B).

NEC 376.23(A) [NEC 378.23(A)] - Deflected Insulated Conductors (Insulated Conductors) - When conductors are deflected (the required bending of the conductor) in a metal [nonmetallic] wireway whether at the point of entry or exit, or where the direction of the deflection exceeds *30 degrees* (30°) the width of the metal [nonmetallic] wireway is determined per Table (NEC) 312.6(A) [See Figure 366.58(A)]. In addition to NEC 376.22(A), 376.56(A) and 378.22, NEC 376.23(A) and 378.23(A) also requires metal [nonmetallic] wireways to be sized according to the deflection of insulated conductors.

NEC 376.23(B) [NEC 378.23(B)] - Metal [Nonmetallic] Wireways Used as Pull Boxes (Insulated Conductors) - Where insulated conductors 4 AWG or larger are *pulled through* a wireway, the distance between raceway and cable entries enclosing the same conductor shall not be less than required in NEC 314.28(A)(1) for straight pulls and NEC 314.28(A)(2) for angle pulls [See Figure 366.58(B)]. When transposing cable size into raceway size, the minimum trade size raceway required for the number and size of conductors in the cable shall be used (See question No. 52. of Article 314).

NEC 376.30 [NEC 378.30] - Securing and Supporting - Metal [Nonmetallic] wireways shall be supported in accordance with NEC 376.30 (A) and (B) [and NEC 378.30(A) and (B)].

NEC 376.56(A) [NEC 378.56] - Splices and Taps - Splices and taps are permitted if they are accessible. The total cross-sectional area of all splice, tap and other conductors sharing a common cross section must not fill (occupy) more than *75 percent* (.75) of the metal [nonmetallic] wireway interior cross-sectional area at that point. To minimize the *75 percent* limitation, splice or tap conductors are often staggered. See Figures 376.56 and 366.1(b) and (c).

WIREWAYS

(At that point)-Where splice, tap and other conductors share a common cross section

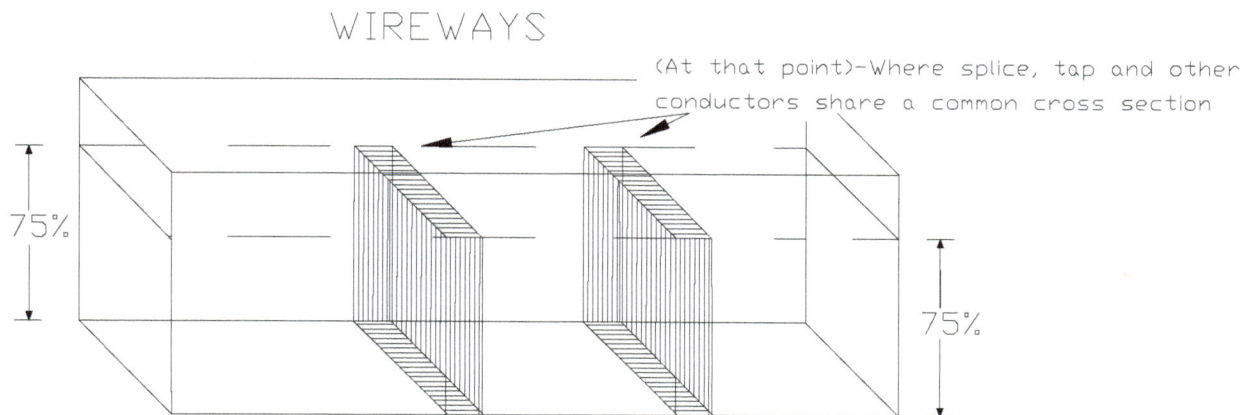

75%

75%

Splice, tap and other conductors sharing a common cross section cannot occupy no more than 75% of the cross-sectional area of the wireway at that point unlike an auxiliary gutter which expands the 75% limitation to that of the entire gutter space where such conductors are permitted [see Figure 366.56(A)].

Figure 376.56

APPLYING THE NEC

The methods used to determine the internal use of an auxiliary gutter or wireway are basically the same and only requires a few short steps. NEC references such as Tables 5, 5A and 8 of Chapter 9 which are primarily used to calculate raceway sizes will be used to calculate the total cross-sectional area of all conductors that are installed in either auxiliary gutters or wireways.

GUTTER and WIREWAY INTERIOR CROSS-SECTIONAL AREA

When determining the *20 percent* limitations of an auxiliary gutter [NEC 366.22(A)] or wireway [NEC 376.22(A) and 378.22], the installed conductor's *at any cross-section* must be totaled *per individual conductor's cross-sectional area* as referenced in Tables 5, 5A and 8 of Chapter 9. Refer to the following procedures.

1. (a) When an installation does not involve splice or tap conductors, refer to Figure 366.1(a).

At any cross section - Identify each set of conductors sharing a common cross section. Total the cross-sectional area of those conductors based upon each individual conductor's size, insulation type and the number of conductors. Select the common cross section containing the *larger/largest* conductor cross-sectional area.

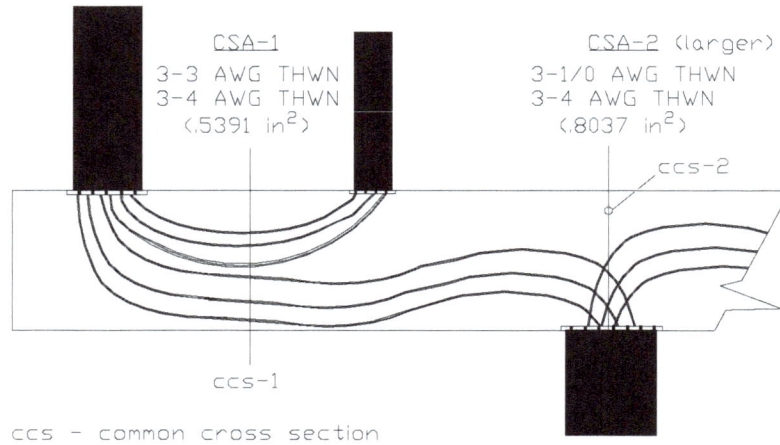

Figure 366.1(a) - Common cross sections-no splice or tap conductors

(b) When an installation involves conductors along with splice and/or tap conductors, refer to Figure 366.1(b). Figure 366.1(b) could represent the use of split bolt connectors or other bulky type splice/tap components that utilizes excessive space.

At any cross section - Identify each set of conductors sharing a common cross section *before* (b) and *after* (a) the splice or tap connections. Total the cross-sectional area of those conductors based upon each individual conductor's size, insulation type and the number of conductors. Select the common cross section containing the *larger/largest* conductor cross-sectional area.

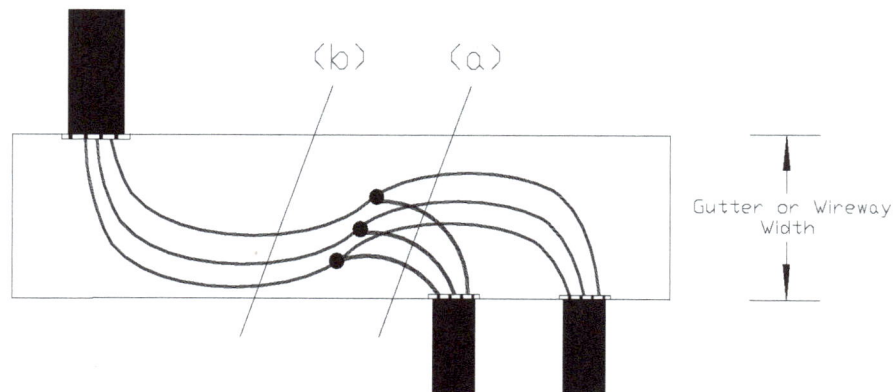

Figure 366.1(b) - Common cross sections with staggered splice and/or tap conductors

Slightly *staggering* such conductors where connected - minimize the area taken up at any cross section thus preventing splice and/or tap conductors from occupying a common cross section - thereby minimizing the 75 percent limitation per NEC 376.56(A) and 378.56 for wireways. For an auxiliary gutter, the limitation is based upon the entire cross-sectional area of the enclosure.

Where splice and/or tap conductors occupy a *common cross section (without staggering) - the total cross-sectional area of such conductors are added together - to gather the minimum size auxiliary gutter or wireway (enclosure) needed without having knowledge of the bulk of the type (splice/tap) connectors or components being used which will further contribute to the 75 per limitation. See Figure 366.1(c).

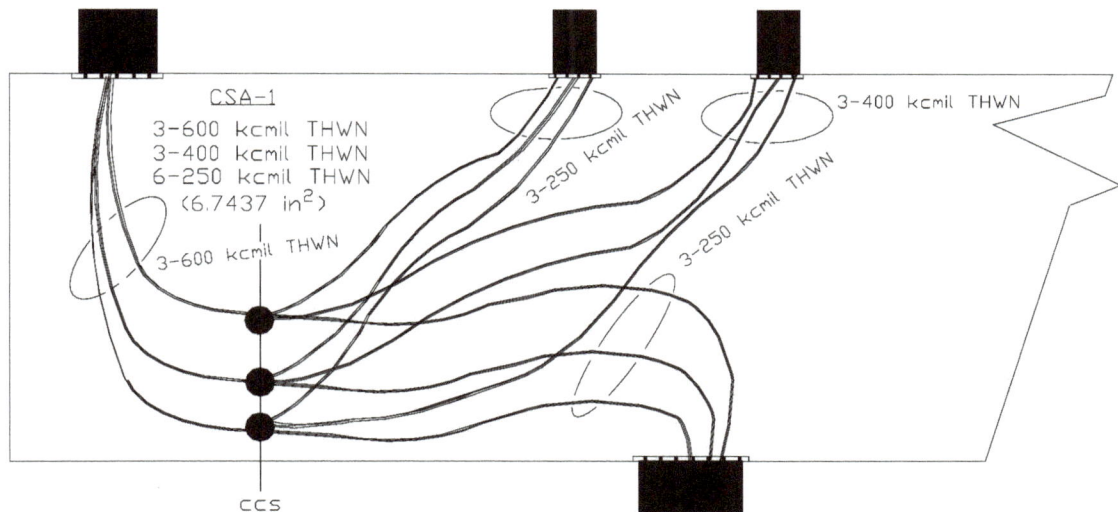

Figure 366.1(c) - Common cross section with unstaggered splice and/or tap conductors

*Where two or more common cross sections exist, consider the larger or largest combinations of splice and/or tap conductors.

2. Multiply the **larger/largest conductor cross-sectional area by 5 to determine the **required cross-sectional area** of the enclosure. Refer to the applicable formulas.

The value *five* (5) is used because the total cross-sectional area of installed conductors is limited to *20 percent* of the total cross-sectional area (100 percent) of an auxiliary gutter or wireway that is, 100 percent (or 1) divided by 20 percent (or .20) is equal to 5.

Required Cross-sectional Area (minimum) = Conductor's total cross-sectional area x **5**

If desired, use the following alternate formula to determine the required cross-sectional area of the enclosure based on the *20 percent* conductor fill limit of the enclosure:

Required Cross-sectional Area (minimum) = $\frac{\text{Conductor's total cross-sectional area}}{.2}$

**The larger/largest (common) conductor cross-sectional area is only applied because the reference requirements states *at any cross-section* therefore, all common areas need not be considered. The largest conductor cross-sectional area can be used to determine the required cross-sectional area of the auxiliary gutter or wireway.

3. Select an auxiliary gutter or wireway with a cross-sectional area (csa) that exceeds the calculated value of the **required cross-sectional area**. See Columns **1** or **2** of ensuing TABLE.

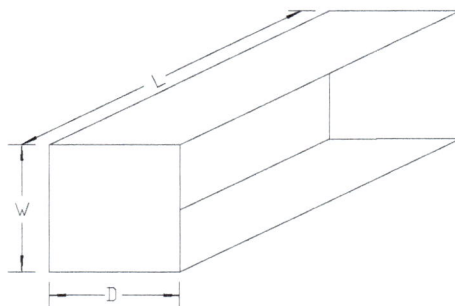

TABLE - GUTTER and WIREWAY STANDARD SIZES (TYPICAL)

Size (in) [D x W]	1	2	3	4
2.5 x 2.5	6.25	1.25	4.6875	2 AWG and less
3 x 3	9.00	1.80	6.75	3 AWG and less
4 x 4	16.00	3.20	12.00	4/0 AWG and less
4 x 6	24.00	4.80	18.00	250 kcmil and less
6 x 6	36.00	7.20	27.00	500 kcmil and less
8 x 8	64.00	12.80	48.00	900 kcmil and less
10 x 10	100.00	20.00	75.00	1250 kcmil and less
12 x 6	72.00	14.40	54.00	500 kcmil and less
12 x 12	144.00	28.80	108.00	2000 kcmil and less

1 - Cross-sectional area (in^2) (at 100%)　　2 - Cross-sectional area (in^2) (at 20%)
3 - Cross-sectional area (in^2) (at 75%)　　4 - Maximum deflected conductor(s)

4. When **step 3** is applied, determine whether the selected size auxiliary gutter or wireway width is adequate per conductor deflection (Column **4** of Table) per NEC and Table 312.6(A) [See Figure 366.58(A)]. When an auxiliary gutter or wireway is used as a pull box, maintain the minimum distance between raceway entries per NEC 366.58(B), 376.23(B) or 378.23(B) according to NEC 328.28(A)(1) or (2) [See Figure 366.58(B)].

366.22(A) - Sheet Metal Auxiliary Gutters *and* **366.22(B)** - Nonmetallic Auxiliary Gutters (Number of Conductors) (1. - 5.) [5]

1. An auxiliary gutter will contain the following copper conductors:

15 - 8 AWG TW	8 - 1 AWG THW	11 - 6 AWG THHN
7 - 2/0 AWG XHHW	10 - 4 AWG THWN	4 - 250 kcmil THHW

Determine the minimum size gutter needed to enclose the conductors.

According to NEC 366.22(A), the total cross-sectional area of all enclosed conductors at any cross-section of the gutter must not exceed 20 percent (.20) of the gutter's interior cross-sectional area.

DETERMINE TOTAL CROSS-SECTIONAL AREA OF CONDUCTORS

Table 5, Chapter 9

8 AWG TW	$.0437$ in^2	x	15	=	$.6555$ in^2
6 AWG THHN	$.0507$ in^2	x	11	=	$.5577$ in^2
4 AWG THWN	$.0824$ in^2	x	10	=	$.8240$ in^2
1 AWG THW	$.1901$ in^2	x	8	=	1.5208 in^2
2/0 AWG XHHW	$.2190$ in^2	x	7	=	1.5330 in^2
250 kcmil THHW	$.4596$ in^2	x	4	=	$\underline{1.8384 \text{ in}^2}$
					6.9294 in^2 (Total)

DETERMINE MINIMUM CROSS-SECTIONAL AREA OF GUTTER

$$6.9294 \text{ in}^2/.20 = 34.65 \text{ in}^2 \textit{ or } 6.9294 \text{ in}^2 \text{ x } 5 = 34.65 \text{ in}^2$$

MINIMUM SIZE GUTTER REQUIRED

A 6" x 6" auxiliary gutter which has a cross-sectional area of 36 in^2 is the *minimum* size required. Although not mentioned, if the installation of these conductors required being deflected greater than 30 degrees, based on the largest conductor (250 kcmil), per Table 312.6(A), the minimum width of the gutter is required to be 4½", where in this situation the minimum width is exceeded by use of the 6" (length) x 6" (width) auxiliary gutter.

2. Refer to Figure 366.1(a). If the illustrated enclosure is an auxiliary gutter, what size gutter is required?

Again, according to NEC 366.22(A), the total cross-sectional area of all enclosed conductors at *any* cross-section of the gutter must not exceed 20 percent (.20) of the gutter's interior cross-sectional area. However, in this situation the phase "at any cross section" reference the two common cross sections identified. Because the referenced cross sections are occupied by conductors having different cross-sectional areas, only the larger cross-sectional area need to be applied when determining the required size of the auxiliary gutter. For certain, if the larger conductor's cross-sectional area is either the same or less than the 20 percent requirement of NEC366.22(A), the smaller conductor's cross-sectional area will be far less than the requirement based on the auxiliary gutter being sized per larger conductor's cross-sectional area. Observe,

$$.8037 \text{ in}^2/.20 = 4.0185 \text{ in}^2$$

According to the calculation, as a minimum, a 4" x 6" standard size gutter will be required that has a cross-sectional area of 24 in^2 and thus will also satisfy the requirements of NEC 366.58(A) (see Columnss **1**, **2**, and **4** of TABLE). At 4.0185 in^2 the largest conductor's cross-sectional area will only occupy 16.74 percent (4.0185 in^2/24 in^2) of the interior cross-sectional area of the gutter which is less than the 20 percent requirement. As for the smaller conductor's cross-sectional area, (.5391 in^2/.20 = 2.6955 in^2) only 11.23 percent (2.6955 in^2/24 in^2) of the interior cross-sectional area of the selected gutter will be occupied which is much less than 16.74 percent and far less than the 20 percent requirement. For such reasons, only the larger or largest common cross section where applicable, is applied.

If conduit sizes were given in Figure 366.1(a), the gutter could possibly be sized as a pull box. See question No. 10.

3. Refer to Figure 366.1(b). Assume the enclosure is an auxiliary gutter where the conductors (copper) *before* the tap connections are 3/0 AWG THW and the conductors (copper) *after* the tap connections are 1/0 AWG and 1 AWG THWN. Determine the minimum size gutter required.

To determine the gutter size needed for this installation, calculate the total cross-sectional areas of the conductors occupying the common cross sections of the gutter *before* and *after* the tap connections are made.

Cross-Sectional Area of Conductors *before* Tap Connections

In this situation the only conductors to consider are the 3-3/0 AWG conductors. Per Table 5 of Chapter 9 the cross-sectional area of the 3/0 AWG THW conductors is .3117 in^2.

$$.3117 \ in^2 \ x \ 3 = .9351 \ in^2$$

Cross-Sectional Area of Conductors *after* Tap Connections

The only conductors to consider are the 3-1/0 and 3-1 AWG THWN conductors. Per Table 5 of Chapter 9, the cross-sectional areas of the 1/0 AWG and 1 AWG THWN are .1855 in^2 and .2624 in^2, respectively. Now the conductor's total cross-sectional area can be determined.

$$
\begin{aligned}
(1/0 \ AWG \ THWN) - .1855 \ in^2 \ x \ 3 &= .5565 \ in^2 \\
(1 \ AWG \ THWN) \quad - .1562 \ in^2 \ x \ 3 &= \underline{.4686 \ in^2} \\
(Total) \quad &\ \ 1.0251 \ in^2
\end{aligned}
$$

Because the total cross-sectional area of the conductors *after* the tap connections is the larger of the two, it will be used to size the gutter.

$$1.0251 \ in^2/.20 = 5.1255 \ in^2$$

The cross-sectional area of the gutter must exceed 5.1255 in^2 which requires a 2.5" x 2.5" auxiliary gutter. The cross-sectional area of a 2.5" x 2.5" auxiliary gutter is 6.25 in^2. Although the gutter was sized based on the larger conductor's cross-sectional area, the deflection of the conductors still has to be considered per NEC 366.58(A). Because the 3/0 AWG conductors are the largest of the conductors to be deflected, according to Table 312.6(A), the width of the gutter is required to be a minimum of 4" at one wire per terminal to accommodate the 3/0 AWG conductors. Therefore, a 4" x 4" gutter must be used instead of a 2.5" x 2.5" gutter.

If the tap connections were not spaced apart (unstaggered) and all tap connections were made in the same cross-section of the gutter, the cross-sectional area of all conductors would have to be considered to ensure the 20 percent limitation is not exceeded, take a look.

Cross-Sectional Area of *all* Conductors

$$.9351 \text{ in}^2 + 1.0251 \text{ in}^2 = 1.9602 \text{ in}^2$$
$$1.9602 \text{ in}^2/.20 = 9.8010 \text{ in}^2$$

To stay within the allowable 20 percent (.20) fill of an auxiliary gutter, a 4" x 4" (16 in^2) gutter would still be required if the tap connections were made in the same cross-section of the gutter. However, just remember that the gutter was initially required to be 2.5" x 2.5" but required being upgraded based upon the deflected conductors.

4. How many 4/0 AWG THWN-2 conductors are allowed to be installed in a 2.5" x 2.5" gutter, a 4" x 4" gutter and a 6" x 6" gutter?

Because the allowable fill of an auxiliary gutter is limited to 20 percent of the gutter's interior cross-sectional area, determine the allowable area first.

$$\begin{array}{ll}
\text{2.5 in. x 2.5 in.} = 6.25 \text{ in}^2 & \text{6.25 in}^2 \text{ x .20} = 1.25 \text{ in}^2 \\
\text{4 in. x 4 in.} = 16 \text{ in}^2 & \text{16 in}^2 \text{ x .20} = 3.20 \text{ in}^2 \\
\text{6 in. x 6 in.} = 36 \text{ in}^2 & \text{36 in}^2 \text{ x .20} = 7.20 \text{ in}^2
\end{array}$$

The total cross-sectional area of the 4/0 AWG THWN-2 conductors being installed in the,

2.5 in. x 2.5 in. auxiliary gutter cannot exceed 1.25 in^2
4 in. x 4 in. auxiliary gutter cannot exceed 3.20 in^2
6 in. x 6 in. auxiliary gutter cannot exceed 7.20 in^2

Per **Table 5, Chapter 9**, a 4/0 AWG THWN-2 has a cross-sectional area of .3237 in^2. Dividing this value into the values of the allowable cross-sectional areas of each gutter determines how many 4/0 AWG THWN-2 conductors can be installed in each gutter.

2.5 in. x 2.5 in. auxiliary gutter - 1.25 in^2 / .3237 in^2 = 3.86 (rounded down) 3
4 in. x 4 in. auxiliary gutter - 3.20 in^2 / .3237 in^2 = 9.89 (rounded down) 9
6 in. x 6 in. auxiliary gutter - 7.20 in^2 / .3237 in^2 = 22.24 (rounded down) 22

Three (3) - 4/0 AWG conductors can be in installed in the 2.5 in. x 2.5 in. gutter
Nine (9) - 4/0 AWG conductors can be in installed in the 4 in. x 4 in. gutter
Twenty-two (22) - 4/0 AWG conductors can be in installed in the 6 in. x 6 in. gutter

5. A 4" x 4" metal auxiliary gutter contains 16 - 4 AWG and 18 - 3 AWG THWN copper conductors. If the conductors are current-carrying, determine the ampacity of the conductors.

Table 310.15(B)(16) lists the ampacity of a 4 AWG THWN copper conductor as 85A and a 3 AWG THWN copper conductor as 100A. Applying the appropriate adjustment factor of Table

310.15(B)(3)(a) requires a 40 percent (.40) adjustment of each conductor based on a total of 34 current-carrying conductors.

$$4 \text{ AWG THWN} - 85 \times .40 = 34A \quad 3 \text{ AWG THWN} - 100 \times .40 = 40A$$

The new ampacity of each conductor is as shown above.

366.23(A) - Sheet Metal Auxiliary Gutter (Ampacity of Conductors) (6. - 7.) [2]

6. How much current is a ½" x 1½" copper busbar capable of carrying continuously in a sheet metal auxiliary gutter?

According to NEC 366.23(A), current carried continuously in a copper bar in a sheet metal auxiliary gutter must not exceed 1000 amperes / in^2 (1000 amperes per square inch).

The cross-sectional area of the "copper busbar" is .75 in^2 (½ in. x 1½ in.). Therefore,

$$\frac{1000 \text{ amperes}}{\text{in}^2} \times .75 \text{ in}^2 = 750 \text{ amperes}$$

The copper busbar can carry 750 amperes continuously.

7. How much current is a 3/8" x 1¼" aluminum busbar capable of carrying continuously in a sheet metal auxiliary gutter?

According to NEC 366.23(A), current carried continuously in an aluminum bar in a sheet metal auxiliary gutter must not exceed 700 amperes / in^2 (700 amperes per square inch).

The cross-sectional area of the "aluminum busbar" is .469 in^2 (3/8 in. x 1¼ in.). Therefore,

$$\frac{700 \text{ amperes}}{\text{in}^2} \times .469 \text{ in}^2 = 328.3 \text{ amperes}$$

The aluminum busbar can carry 328.3 amperes continuously.

366.23(B) - Nonmetallic Auxiliary Gutters (Ampacity of Conductors)

Unlike metal gutters, when nonmetallic auxiliary gutters contain more than three current-carrying conductors the derating factors of NEC and Table 310.16(B)(3)(a) must be applied. Therefore, if the conductors in question No. 5. were in a nonmetallic auxiliary gutter instead, the results would be the same with the exception of the conductor's derating ampacity occurring after the count of three current-carrying conductors opposed to those exceeding 30.

366.56(A) - Within Gutters (Splices and Taps) (8. - 9.) [2]

8. A 2" rigid steel conduit is mounted on the bottom wall of an auxiliary gutter near the right end of the gutter. Mounted on the top wall near the left end of the gutter are two 1½" electrical metallic tubing (EMT) which are distanced 12" apart. Three 250 kcmil THHN copper conductors are pulled into the gutter through the 2" rigid steel conduit. Three 1/0 AWG THHN copper conductors and 3 - 2/0 AWG THW-2 copper conductors are pushed in separately through the 1½" EMT and tapped to the 250 kcmil conductors. Determine the minimum size auxiliary gutter required for this installation and whether the 75 percent fill limitation for tap conductors is exceeded. Assume all taps occupy the same common area. If needed, sketch the layout of the installation before beginning or refer to Figure 366.1(c) as an illustrated example.

CROSS-SECTIONAL AREA OF TAP CONDUCTORS

Per **Table 5 of Chapter 9**, the cross-sectional area of the 250 kcmil THHN conductors is .3970 in^2 whereas the cross-sectional areas of the 1/0 AWG THHN and 2/0 AWG THW-2 are .1855 in^2 and .2624 in^2, respectively. Because the taps occupy the same common area the conductor's cross-sectional areas are all added together.

$$(250 \text{ AWG THHN}) - .3970 \text{ in}^2 \times 3 = 1.1910 \text{ in}^2$$
$$(1/0 \text{ AWG THHN}) - .1855 \text{ in}^2 \times 3 = .5565 \text{ in}^2$$
$$(2/0 \text{ AWG THW-2}) - .2624 \text{ in}^2 \times 3 = .7872 \text{ in}^2$$
$$(\text{Total}) \quad 2.5347 \text{ in}^2$$

Based on the calculated results, the size gutter needed to house the tap conductors must exceed the value of the following calculation:

$$\frac{2.5347 \text{ in}^2}{.20} = 12.6735 \text{ in}^2$$

At 12.6735 in^2, as a minimum a 4" x 4" gutter, which has a cross-sectional area of 16 in^2 must be used. However, before coming to such conclusion there are other factors to consider.

Because all involved conductors would require being deflected greater than 30 degrees in order to tap the 2/0 and 1/0 AWG conductors to the 250 kcmil conductors, the width of the gutter is required to be sized in accordance with Table 312.6(A) as reference in NEC 366.58(A). Per Table 312.6(A), based on the 250 kcmil conductors - the largest set of conductors, the width of the gutter is required to be a minimum of 4½" based upon one wire per terminal. Requiring a 4½" width will not allow a 4" x 4" gutter to be used but instead will now require the use of a 4" x 6" gutter at 24 in^2. (See TABLE).

Although each set of conductors entered the gutter through different raceways, each set of conductors also terminates in the gutter as well therefore; the possible use of NEC 366.58(B) is eliminated, for this provision requires conductor to *be pulled through an auxiliary gutter*. In order to apply NEC 366.58(B) which reference the use of NEC 314.28(A) the conductors would require being transitioned continuously (without division) from one raceway entry to another.

Having now determined the minimum size auxiliary gutter for this installation, the 75 percent fill limitation will limit the enclosed tap conductors to an area not exceeding 18 in^2 (24 in^2 x .75) which is clearly beyond the calculated cross-sectional area of the tap conductors, 2.5347 in^2. At 2.5347 in^2, the tap conductors will only occupy 10.56 percent of the gutter (2.5347 in^2 / 24 in^2) per calculation.

In concluding, one final thing must be remembered; in an actual installation the illustrated display of either splice or tap connectors as shown in Figures 366.1(B) and (c) are far less in dimensions than the actual bulk of such connectors (split bolts, spice/tap kits, etc.) and the completed work. Therefore, the realization of the 75 percent limitation will not become apparent until an actual situation is encountered.

9. Refer to question No. 3. Does the size of the selected auxiliary gutter allow the 75 percent fill limitation to be maintained?

A 4" x 4" gutter which has a cross-sectional area of 16 in^2 yields at 75 percent a cross-sectional area of 12 in^2. Because each set of identical tap conductors are staggered and maintain separate cross sections, only one set of the three tap conductors need be considered which totals .6534 in^2 (.3117 in^2 + .1855 in^2 + .1562 in^2) in cross-sectional area. At .6534 in^2, each set of tap conductors will only consume approximately 4.1 percent (.6534 in^2/16 in^2) of the gutter's cross-sectional area. Again, such results does not take into consideration the actual size of the adjoining tap connectors which will add more bulk and consume more of the gutter's cross-sectional area.

366.58(A) - Deflected Insulated Conductors (Insulated Conductors)

See question Nos. 1., 3. and 8.

366.58(B) - Auxiliary Gutters Used as Pull Boxes (Insulated Conductors)

10. Refer to Figure 366.1(a). Assume the two top-mounted conduits were sized from left to right at 1¼" and 1" and the bottom-mounted conduit was sized at 1½", what size gutter would be required?

In question No. 2., the gutter was sized according to NEC 366.22(A). Since the attached conduits have now been sized, the gutter can likewise be sized as a pull box per NEC 366.58(B) in accordance with NEC 314.28(A)(2) for *angle* and *u* pulls.

In applying the provisions of NEC 314.28(A)(2), the distance between each raceway entry and the opposite wall shall be not less than six times the largest raceway in a row where the distance is increased for additional entries by the amount of the sum of the diameters of all other raceway entries. Considering the top wall of the gutter where 1¼" and 1" conduits are mounted, the distance between the raceway entries and the opposite wall must be, 1¼" x 6 + 1" = **8½"**.

As for the bottom wall where a 1½" conduit is mounted, the distance between this raceway entry and the opposite wall must be, 1½" x 6 = **9"**.

In compliance with NEC 314.28(A)(2), which requires the maximum distance to be used, the gutter must have a width that is no less than 9". Referring to the TABLE, a 10" x 10" gutter is now required which is a considerable difference from the 4" x 4" gutter required based upon the provisions of NEC 366.22(A). Taking into account the standard size auxiliary gutters listed in the TABLE and the provisions of NEC 314.28(A)(2), a raceway entry or a combination of raceway entries opposite the walls of an auxiliary gutter or wireway exceeding 12" in distance will eliminate the use of such enclosures due to the width restriction of the largest listed standard size enclosure (12" x 12").

366.100(E) - Clearance of Bare Live Parts

11. An auxiliary gutter is in need of being sized to enclose 4 - ¾" x 1¼" copper busbars. The busbars will be horizontally positioned side by side. Determine the minimum dimensions of the auxiliary gutter needed. If needed, refer to Figure 366.100(E).

NEC 366.100(E) requires a 2" minimum clearance between bare current-carrying metal parts and a 1" minimum clearance between bare current-carrying metal parts and any metal surface.

DETERMINE WIDTH OF GUTTER

The width of the busbars is ¾". Minimum spacing required between busbars is 2". Minimum spacing required between busbars and metal walls of gutter is 1".

WIDTH OF GUTTER
1" + ¾" + 2" + ¾" + 2" + ¾" + 2" + ¾" + 1" = 11"

DETERMINE DEPTH OF GUTTER

The height of the busbars is 1¼". Minimum spacing required between busbars and metal walls of gutter is 1".

DEPTH OF GUTTER
1" + 1¼" + 1" = 3¼"

The minimum width (left to right walls) of the gutter must be 11". The minimum depth (top to bottom walls) of the gutter must be 3¼".

376.22(A) - Cross-Sectional Areas of Wireway (Number of Conductors and Ampacity) *and*
378.22 - Number of Conductors (12. - 14.) [3]

12. Determine the minimum size metal wireway required to enclose the following copper conductors: 7-3/0 AWG THWN and 11-10 AWG THW.

Application (as outlined)

Step 1: Since the installation involves no splice or tap conductors,

Refer to **Table 5 of Chapter 9** and select the approximate csa of each dissimilar conductor to determine the total csa of all conductors.

$$(3/0 \text{ AWG THWN}) \quad .2679 \text{ in}^2 \text{ x } 7 = 1.8753 \text{ in}^2$$
$$(10 \text{ AWG THW}) \quad .0243 \text{ in}^2 \text{ x } 11 = \underline{.2673 \text{ in}^2}$$
$$2.1426 \text{ in}^2 \text{ (Total csa)}$$

Step 2: Multiply total csa by 5 or use alternative.

$$2.1426 \text{ in}^2 \text{ x } 5 = 10.713 \text{ or } \frac{2.1426 \text{ in}^2}{.2} = 10.713 \text{ in}^2$$

Step 3: Select the minimum size wireway based on the calculated values.

Referring to the previously given TABLE (WIREWAY AND GUTTER STANDARD SIZES) provided, a 4" x 4" wireway has a cross-sectional area of 16 in². Since the cross-sectional area of this wireway exceeds the calculated value above, 10.713 in² and because there is no conductor deflection involved, the wireway can be used.

Use a "4 x 4" wireway.

13. Refer to Figure 376.22(A)/378.22-13.

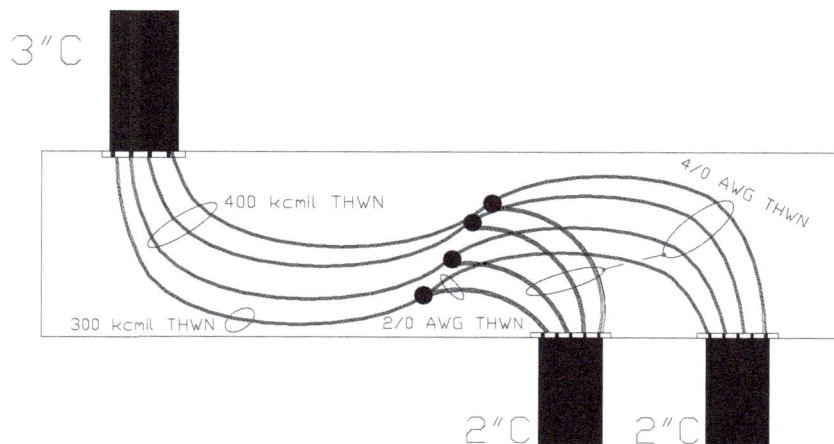

Figure 376.22(A)/378.22-13

What size wireway is required to enclose the tap conductors?

The provisions of NEC 376.22(A) and 378.22 are the same as NEC 366.22(A) which require the sum of the cross-sectional areas of all contained conductors at *any* cross section of a wireway to not exceed 20 percent of the interior cross-sectional area of the wireway. The same procedure used in question No. 3. can be used to determine the size wireway needed for this installation.

To determine the size wireway needed for this installation, calculate the total cross-sectional areas of the conductors occupying the common cross sections of the wireway *before* and *after* the tap connections are made.

Per Table 5 of Chapter 9 (conductor cross-sectional areas)

Cross-Sectional Area of Conductors *before* Tap Connections

$$400 \text{ kcmil THWN - } .5863 \text{ in}^2 \text{ x } 3 = 1.7589 \text{ in}^2$$
$$300 \text{ kcmil THWN - } .4608 \text{ in}^2 \text{ x } 1 = \underline{.4608 \text{ in}^2}$$
$$(\text{Total}) \quad 2.2197 \text{ in}^2$$

Cross-Sectional Area of Conductors *after* Tap Connections

$$4/0 \text{ THWN - } .3237 \text{ in}^2 \text{ x } 6 = 1.9422 \text{ in}^2$$
$$2/0 \text{ THWN - } .2223 \text{ in}^2 \text{ x } 2 = \underline{.4446 \text{ in}^2}$$
$$(\text{Total}) \quad 2.3868 \text{ in}^2$$

Because the total cross-sectional area of the conductors *after* the tap connections is the larger of the two, it will be used to size the wireway.

$$2.3868 \text{ in}^2/.20 = 11.9340 \text{ in}^2 \text{ } or \text{ } 2.3868 \text{ in}^2 \text{ x } 5 = 11.9340 \text{ in}^2$$

The cross-sectional area of the wireway must be no less than 11.9340 in^2 which requires a 4" x 4" wireway. The cross-sectional area of a 4" x 4" wireway is 16 in^2.

14. Refer to Figure 366.1(c). Assume the enclosure is a wireway. Determine the minimum size wireway required.

As earlier discussed and demonstrated, based upon all conductors being unstaggered and connected at a common cross section, the largest conductor cross-sectional area occurs at this cross section. In this arrangement the need for considering the cross-sectional area of conductors *before* and *after* the tap connections can be omitted. By using the total cross-sectional area of all adjoining conductors sharing this common cross section, the minimum size of the required wireway can be determined.

$$\frac{6.7437 \text{ in}^2}{.20} = 33.719 \text{ in}^2$$

Applying the calculated results, at 33.719 in^2, a 6" x 6" wireway which has a cross-sectional area of 36 in^2 is required.

Considering the largest deflected conductors at 600 kcmil conductors, the width of the gutter is required to be sized in accordance with Table 312.6(A) as reference in NEC 376.23(A) and 378.23(A). Per Table 312.6(A), based on the 600 kcmil conductors - the largest set of conductors, the width of the gutter is required to be a minimum of 8" based upon one wire per terminal. Requiring an 8" width will not allow a 6" x 6" gutter to be used but instead will now require the use of an 8" x 8" gutter at 64 in^2 (See TABLE).

376.23(A)/378.23(A) - Deflected Insulated Conductors (Insulated Conductors)

15. Refer to Figure 376.22(A)/378.22-13. What size wireway is required to accommodate the enclosed conductors based upon the provisions of NEC 376.23(A) and 378.23(A)?

Per NEC 376.23(A) and 378.23(A), the width of the gutter is required to be sized in accordance with Table 312.6(A) as reference in NEC 366.58(A). Per Table 312.6(A), based on the 400 kcmil conductors - the largest set of conductors, the width of the wireway is required to be a minimum of 6" based upon one wire per terminal; therefore a 6" x 6" wireway is required.

376/378.23(B) - Metallic/Nonmetallic Wireways Used as Pull Boxes (Insulated Conductors) (16. - 17.) [2]

16. A wireway is being used as a *straight pull box* to route type Z copper conductors ranging from 4 AWG to 1 AWG. From left to right, 1½", 1¼" and 1" raceways are mounted to the top wall of the wireway while 1¼", 1" and two ¾" raceways are mounted to the bottom wall of the wireway. Based on the information provided, what size metal or nonmetallic wireway is required?

In accordance with NEC 376.23(B) and 378.23(B), as it pertains to metal and nonmetallic wireways being used as a pull box, the provisions of NEC 314.28(A)(1) and 314.28(A)(2) are referenced for such straight and angle pull applications. In referencing NEC 314.28(A)(1) for a straight pull application, the only dimension referenced is that of the required length. As a result, the length of the wireway in this situation is required to be no less than eight (8) times the trade size of the largest raceway.

In referring to the largest raceway, which is sized at 1½", the wireway is required to be as a minimum, from the top to bottom wall of the wireway 12" by multiplying 1½" x 8.

17. If the wireway in question No. 16. was being used as an *angle pull box* instead, what size metal or nonmetallic wireway is required?

In applying the provisions of NEC 314.28(A)(2), the distance between each raceway entry and the opposite wall shall be not less than six (6) times the largest raceway in a row where the distance is increased for additional entries by the amount of the sum of the diameters of all other raceway entries. Considering the top wall of the wireway where 1½", 1¼" and 1" raceway is mounted, the distance between the raceway entries and the opposite wall must be,

$$1½" \times 6 + 1¼" + 1" = 11¼"$$

As for the bottom wall where 1¼", 1" and two ¾" raceway is mounted, the distance between the raceway entries and the opposite wall must be,

$$1¼" \times 6 + 1" + ¾" + ¾" = 10"$$

In compliance with NEC 314.28(A)(2), which requires the maximum distance to be used, the wireway must have a width that is no less than 11¼". Referring to the TABLE, a 12" x 12" wireway is required.

376.56(A)/378.56 - Splices and Taps

18. Refer to Figure 376.22(A)/378.22-13. Calculate the conductor's cross-sectional area at each tap connection. Determine whether each tap connection is within the allowed limit per point of tap as required for a wireway. Use the wireway sized in question No. 15.

In observing the four tap connections, it can be said that they are all staggered and have separate connection points. Also, three of the tap connections are identical consisting of one 400 kcmil THWN conductor and two 4/0 AWG THWN conductors while the remaining tap connection consist of one 300 kcmil THWN conductor and two 2/0 AWG THWN conductors. Because three of the tap connections are identical only one calculation is required along with the remaining tap connection. The cross-sectional area of the three identical tap connections yields,

$$(400 \text{ kcmil THWN}) \ .5863 \text{ in}^2 \text{ x } 1 = \ .5863 \text{ in}^2$$
$$(4/0 \text{ AWG THWN}) \ .3237 \text{ in}^2 \text{ x } 2 = \ \underline{.6474 \text{ in}^2}$$
$$1.2337 \text{ in}^2 \text{ (Total csa)}$$

The cross-sectional area of the remaining tap connection yields,

$$(300 \text{ kcmil THWN}) \ .4608 \text{ in}^2 \text{ x } 1 = \ .4608 \text{ in}^2$$
$$(2/0 \text{ AWG THWN}) \ .2223 \text{ in}^2 \text{ x } 2 = \ \underline{.4446 \text{ in}^2}$$
$$.9054 \text{ in}^2 \text{ (Total csa)}$$

Based on the above calculations, each set of totaled values (1.2337 in² and .9054 in²) falls far below the requirements of NEC 376.56(A) and 378.56 which states that the wireway must not be filled to more than 75 percent of its area (27 in² [36 in² x .75]) at that point (where the tap connections take place). Even if the tap connections were unstaggered and made at a common cross section such would only occupy 4.6065 in² (1.2337 in² x 3 + .9054 in²).

If this was the case, only 12.80 percent (4.6065 in²/36 in²) of the wireway at the point of connections would be filled opposed to the permitted 75 percent limitation (27in²/36 in²). However, as mentioned with that for auxiliary gutters, in an actual installation the illustrated display of the tap connectors as shown in Figure 376.22(A)/378.22-13 is far less in dimensions than the actual bulk of such connectors (split bolts, spice/tap kits, etc.) and the completed work. Again, the realization of the 75 percent limitation will not become apparent until an actual situation is encountered. Questions such as these only intended purpose is for the practice and formalization of performing (electrical and mathematical) calculations.

ARTICLE 368 - Busways

368.17(B), *Exception* - Reduction in Ampacity Size of Busway (Overcurrent Protection)

1. A busway in an industrial establishment is protected by a 1000A overcurrent protective device. In various locations of the establishment, sections of the busway not exceeding 43 feet in lengths are reduced in ampacity to supply certain loads. Determine the minimum ampacity ratings of the reduced sections of the busway.

According to the *Exception* to NEC 368.17(B), in an industrial establishment where a smaller size busway is utilized and the length of the smaller busway does not exceed 50 feet, the ampacity of the busway can be reduced to one-third (1/3 or .333) the rating or setting of the overcurrent device next back on the line, where in this case is the 1000A overcurrent protective device. Therefore,

$$1000A \times .333 = 333A$$

which requires the ampacity of the smaller busway to be no less than 333A. In such case, if the smaller busway was rated less than 333A, overcurrent protection would be required at the point of reduction.

ARTICLE 370 - Cablebus

370.4(B) - Ampacity of Conductors,
370.4(C) - Size and Number of Conductors (Conductors) *and*
370.5 - Overcurrent Protection (1. - 2.) [2]

1. A cablebus assembly is being used to carry a 795A continuous load. The assembly will be protected by a 1000A overcurrent device at 575 volts. What size THW copper conductors are required to be used for the cablebus assembly?

In reference to NEC 370.4(B) the ampacity of the conductors will be based on Table 310.15(B)(17) because of the referenced use of type THW insulation. NEC 370.5 states that the cablebus conductors must be protected against overcurrent in accordance with NEC 240.4. Because the overcurrent device is rated above 800 amps, the ampacity of the conductors must be either equal to or greater than the rating of the overcurrent device according to NEC 240.4(C). In accordance with Table 310.15(B)(17), 1250 kcmil THW copper conductors are required which have a rated ampacity of 1065 amps.

2. Assume the cablebus assembly in question No. 1. was being used to supply a 2400V system. What size copper or aluminum conductors at 90°C would be required?

According to NEC 370.4(B) when a cablebus assembly is rated over 600V the ampacity of the conductors in a cablebus must be determined based on the listed ampacities of either Table 310.60(C)(69) or Table 310.60(C)(70). When both tables are referenced, Table 310.60(C)(69) reference the use of insulated single copper conductors at 90°C and 105°C while Table 310.60(C)(70) reference the use of insulated aluminum conductors with likewise temperature ratings.

Conductor's ampacity (2001 – 5000 Volts)
Per Table 310.60(C)(69) - if copper conductors were used, a 1000 kcmil conductor @ 90°C (1075A) would be required and per Table 310.60(C)(70) - if aluminum conductors were used, a 1500 kcmil conductor @ 90°C (1105A) would be required.

In accordance with the *Exception* to NEC 370.5
As for the use of an overcurrent device, the *Exception* to NEC 370.5 reference the use of NEC 240.100 and NEC 240.101 for installations over 600V. According to NEC 240.101(A), if the device consisted of fuses the continuous ampere rating of the fuses would not be allowed to exceed three times the ampacity of the conductors. If copper (1075A x 3 = 3225A) or aluminum (1105A x 3 = 3315A) conductors were used, the overcurrent device would be limited to 3000A. If the device consisted of a circuit breaker with a long-trip element setting or electronically actuated fuses with a minimum trip setting, the ampere rating of either device(s) would not be allowed to exceed six times the ampacity of the conductors. If copper (1075A x 6 = 6450A) or aluminum (1105A x 6 = 6630A) conductors were used the overcurrent device would be limited to 6000A. As you can see in either situation the 1000A overcurrent device where used in an installation rated less than 600V falls far below that which is required for an installation exceeding 600V.

ARTICLE 384 - Strut-Type Channel Raceway

384.22 - Number of Conductors (1. - 2.) [2]

1. A 1½" x 1½" strut-type channel raceway is being installed to enclose 12 AWG THW copper conductors to supply a variety of overhead light fixtures in a large manufacturing facility. How many 12 AWG conductors can be enclosed in the raceway if external joiners will be used to connect sectional parts of the raceway?

According to the footnote to Table 384.22, when external joiners are used the allowable fill of the raceway's total cross-sectional area is 40 percent. For a 1½" x 1½" strut-type channel raceway, the allowable cross-sectional area at 40 percent is .731 in^2. Referring to Table 5 of Chapter 9, the cross-sectional area for a 12 AWG THW conductor is .0181in^2. The number of conductors can be derived by dividing the allowable cross-sectional area of the raceway by the cross-sectional area of a 12 AWG conductor.

$$.731 \text{ in}^2 / .0181 \text{ in}^2 = 40 \text{ conductors (rounded down)}$$

2. Strut-type channel raceway will be used to enclose the following conductor combinations:

3 - 4 AWG XHHW   3 - 8 AWG THWN   4 - 6 AWG THW   5 - 12 AWG THHW

Determine the minimum size raceway required if internal joiners are used.

Refer to Table 5 of Chapter 9 to obtain the cross-sectional area of each conductor.

$$
\begin{aligned}
12 \text{ AWG THHW} &= .0181 \text{ in}^2 \text{ x } 5 = .0905 \text{ in}^2 \\
8 \text{ AWG THWN} &= .0366 \text{ in}^2 \text{ x } 3 = .1098 \text{ in}^2 \\
6 \text{ AWG THW} &= .0726 \text{ in}^2 \text{ x } 4 = .2904 \text{ in}^2 \\
4 \text{ AWG XHHW} &= .0814 \text{ in}^2 \text{ x } 3 = \underline{.2442 \text{ in}^2} \\
&\phantom{= .0814 \text{ in}^2 \text{ x } 3 = } .7349 \text{ in}^2
\end{aligned}
$$

Considering the 25 percent fill column of Table 384.22 per footnote, the minimum size raceway that can be used is 1-5/8" x 2-7/16" which has a .792 in^2 allowable fill area at 25 percent. This size raceway will accommodate the total cross-sectional area of all conductors.

ARTICLE 390 - Under floor Raceways

390.6 - Maximum Number of Conductors in Raceway

1. The interior dimensions of an underfloor raceway measures 5" x 2¾". If the combined cross-sectional area of the conductors filling the raceway is 6.6819 in^2, can the raceway be used?

According to NEC 390.6, the combined cross-sectional area of all conductors enclosed in underfloor raceway cannot exceed 40 percent of the raceway's interior cross-sectional area. Considering the total cross-sectional area of the raceway,

$$5" \text{ x } 2\frac{3}{4}" = 13.75 \text{ in}^2$$

40 percent of the raceway's interior cross-sectional area is equal to 5.5 in^2 (13.75in^2 x .40). Since the conductor's combined cross-sectional area exceeds 5.5in^2, the raceway cannot be used unless the conductor's combined cross-sectional area is reduced 1.1819in^2 (6.6819 in^2 - 5.5 in^2) or less.

Article 392 of the National Electrical Code underwent major changes in the 2011 edition. Although the scope and contents of this Article remained the same, numerical, format and added changes has caused the entire Article to take on a different look. Therefore, beware of the challenges that these new changes may present.

ARTICLE 392 - Cable Trays

Routing conductors from one location to another is not always available or practicable by means of raceway (conduit or tubing). Unlike raceway, when installed in cable tray, conductors are not totally enclosed therefore resulting in this method of conductor routing not being recognized by the National Electrical Code as raceway. Because cable tray is usually installed in high ceiling areas or isolated locations, conductors installed in cable tray are considered protected from physical damage.

392.22(A)(1)(a) - Ladder or Ventilated Trough Cable Trays Containing Any Mixture of Cables [Cables 4/0 AWG or larger] (Number of Multiconductor Cables, Rated 2000 Volts or Less, in Cable Trays) (1. - 2.) [2]

Note: Where applicable or not stated, as it pertains to question Nos. 1. - 13., all conductor terminations will be rated for 75°C.

1. Refer to Figure 392.22(A)(1)(a)-1. Seven individual Type TC cables contain four 300 kcmil THWN copper conductors. The cables are installed side by side in a single layer unspaced. The diameter of each cable is 1.93" and all cable conductors are current-carrying. If a steel ladder cable tray is used, determine the minimum width of the tray and the ampacity of each cable's conductors.

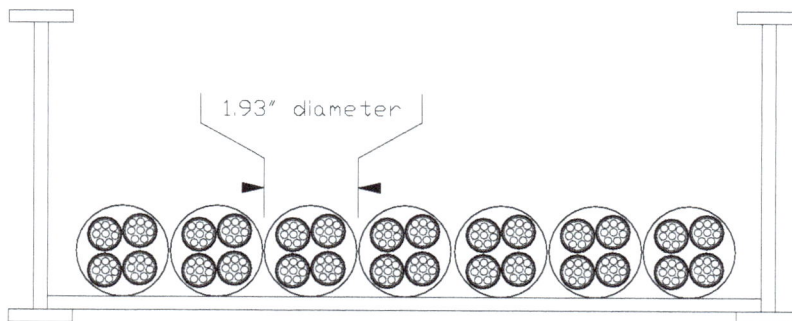

1.93" diameter

Figure 392.22(A)(1)(a)-1

WIDTH OF CABLE TRAY

Although not specifically stated, when multiconductor cables in sizes 4/0 AWG or larger are installed *without spacing* in cable tray, NEC 392.22(A)(1)(a) is referenced. According to NEC 392.22(A)(1)(a), the sum of the diameters of all cables when calculated must not exceed the width of the cable tray. If so, the minimum width of the cable tray must be, 13.51" (7 x 1.93") which will require a ladder tray having a standard width of 16", see Table 392.22(A).

AMPACITY OF CONDUCTORS

Although not specifically worded as so, NEC 392.80(A)(1) reference the use of NEC 392.80(A)(1)(a) for determining the ampacity of multiconductor cables installed *without spacing* in cable tray. Per NEC 392.80(A)(1)(a), when a multiconductor cable contains more than three current-carrying conductors, the derating factors of NEC 310.15(B)(3)(a) must be applied and derating only applies to the number of current-carrying conductors in each cable which is four and not to the number of conductors in the cable tray, which totals 28 conductors.

Referring to Table 310.15(B)(3)(a) an 80 percent adjustment factor is required for four current-carrying conductors. The ampacity of a 300 kcmil THWN copper conductor is 285 amps (Table 310.15(B)(16). After applying the adjustment factor the new ampacity of each conductor is limited to 228 amps (285 amps x .80). NEC 310.15(B)(3)(a)(1) reference the use of NEC 392.80.

2. Refer to Figure 392.22(A)(1)(a)-2. Consider the cables in question No. 1., if the cables now contain three 300 kcmil THWN copper conductors, are installed according to NEC 392.80(A)(1)(c); where the spacing width between cables are the same as the diameter of a single cable and are exposed to an ambient temperature of 121°F. Based on the given conditions, determine the minimum width of the steel ladder cable tray and the ampacity of each cable's conductors.

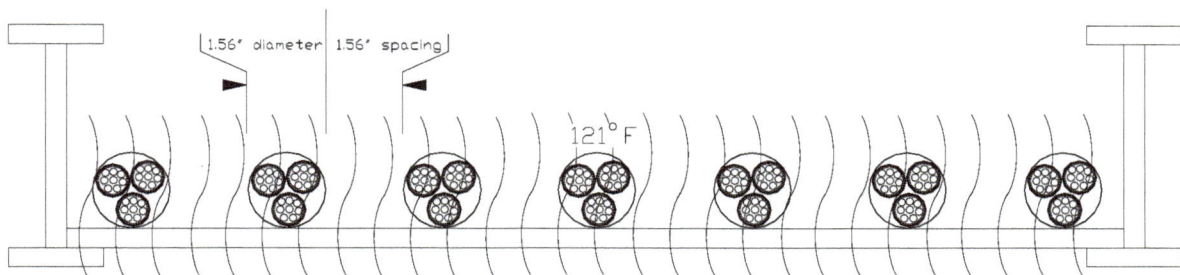

Figure 392.22(A)(1)(a)-2

WIDTH OF CABLE TRAY

According NEC 392.22(A)(1)(a), where the cable ampacity is determined per NEC 392.80(A)(1)(c), the cable tray width shall not be less than the sum of the diameters of the cables and the sum of the required spacing width between the cables. Being the case, the minimum width of the cable tray must be, 10.92" (7 x 1.56" - *sum of the diameters of the cables*) + 9.36" (6 x 1.56" - *sum of the required spacing width between the cables*) totaling 20.28" which will require a ladder tray having a standard width of 24", see Table 392.22(A).

AMPACITY OF CONDUCTORS

When multiconductor cable are installed as shown in Figure 392.22(A)(1)(a)-2 per NEC 392.80(A)(1)(c), the ampacity of the cable's conductors must be determined based on the general equation given in NEC 310.15(C) which requires engineering supervision. However, based on the recommendation provided in NEC 392.80(A)(1)(c) Informational Note, the ampacity of the conductors are generally determined per Table B.310.15(B)(2)(3) of Informative Annex B. Referring to Table B.310.15(B)(2)(3) [40°C (104°F)], the allowed ampacity of the 300 kcmil copper conductors at 75°C (167°F) is 306A. Because the conductors will be exposed to an ambient temperature of 121°F, the footnote to Table B.310.15(B)(2)(3) reference the use of NEC 310.15(B)(2) for the ampacity correction factors when the ambient temperature is other than 40°C (104°C). Applying the given equation in NEC 310.15(B)(2), the ampacity corrected for the ambient temperature is,

$$I' = 306A \times \left(\sqrt{\frac{167-121}{167-104}} = \sqrt{\frac{46}{63}} = \sqrt{.7302} = .8545 \right)$$

$$I' = 306A \times .8545 = 261.48A$$

In comparison, if the relative correction factor of Table 310.15(B)(2)(b) [40°C (104°F)] was applied (.85), the conductor's allowed ampacity is reduced to 260.10A (306A x .85).

392.22(A)(1)(b) - Ladder or Ventilated Trough Cable Trays Containing Any Mixture of Cables [Cables smaller than 4/0 AWG] (Number of Multiconductor Cables, Rated 2000 Volts or Less, in Cable Trays)

3. A 96' trough cable tray with a continuous solid non-ventilated cover is being used to route the following combinations of Type TC cables:

4 - 1/0-3 aluminum XHHW
5 - 2/0-3 aluminum THWN
3 - 3/0-3 aluminum THW

All conductors are rated for 75°C and the cross-sectional area of each individual cable from smallest to largest is .602 in², .734 in², 1.0286 in². Determine the minimum width of the cable tray and the ampacity of each cable's conductors.

WIDTH OF CABLE TRAY

Unlike the provisions of NEC 392.22(A)(1)(a), (which only requires the width of a cable tray to be determined based on either the diameter of all cables *or* the diameter of all cables and the spacing width between cables) NEC 392.22(A)(1)(b) requires the width of the cable tray to be based on the maximum allowable fill areas given in Column 1 of Table 392.22(A) that exceeds the total cross-sectional area of the conductors. The total cross-sectional area (csa) of all cable is as calculated.

$$4\text{-}1/0 \text{ AWG} \quad .602 \text{ in}^2 \text{ x } 4 \quad = 2.408 \text{ in}^2$$
$$5\text{-}2/0 \text{ AWG} \quad .734 \text{ in}^2 \text{ x } 5 \quad = 3.670 \text{ in}^2$$
$$3\text{-}3/0 \text{ AWG} \quad 1.0286 \text{ in}^2 \text{ x } 3 = \underline{3.086 \text{ in}^2}$$
$$9.164 \text{ in}^2 \text{ (total csa cable)}$$

In comparison to the total cross-sectional area of the cable and column 1 of Table 392.22(A), the cable tray is required to have a maximum allowable fill area of 9.5 in² and a corresponding width of 8" as shown in the column preceding column 1.

Table 392.22(A) lists the standard (inside) widths of cable trays ranging from 2 to 36 inches. **Column 1** of Table 392.22(A) lists the allowed (usable) cross-sectional areas (csa) for multiconductor cables in ladder and ventilated trough cable tray which corresponds to approximately 40 percent (.40) of the cable tray's total cross-sectional area. The allowed cross-sectional areas listed in **Column 1** were derived based on the given widths of the cable tray and a standard manufactured cable tray depth of 3 inches. For example, the width of the first listed cable tray (2 in.) *times* the standard manufactured depth (3 in.) *times* 40 percent (.40) yields,

$$2 \text{ in. x } 3 \text{ in. x } .40 = 2.4 \text{ in}^2$$

Based on the calculated results, the allowed cross-sectional area is approximate to the listed 2.5 in² in **Column 1** for multiconductor cables. However, just as the 2.4 in² calculated value at 40 percent was rounded up to the listed 2.5 in² which amounts to an allowed cross-sectional area slightly above 40 percent (2.5 in²/6 in² = .4167 or 41.67 percent); if calculated, the results of all remaining cross-sectional area values listed in **Column 1** would yield an allowed cross-sectional area percentage value that is approximately 40 percent.

AMPACITY OF CONDUCTORS

According to NEC 392.80(A)(1)(b), the ampacity of the conductors cannot exceed 95 percent (.95) of the ampacity listed in Tables 310.15(B)(16) or 310.15(B)(18). If the conductors were required to operate in an environment where the ambient temperature and the number of current-carrying conductors were exceeded, the ampacity of the conductors would require being reduced even more.

Referring to Table 310.15(B)(16) based upon the temperature rating of the conductors, the required ampacities of the aluminum conductors rated for 75°C per NEC 392.80(A)(1)(b) are,

$$1/0 \text{ AWG - } 120A \text{ x } .95 = 114A$$
$$2/0 \text{ AWG - } 135A \text{ x } .95 = 128.25A$$
$$3/0 \text{ AWG - } 155A \text{ x } .95 = 147.25A$$

392.22(A)(1)(c) - Ladder or Ventilated Trough Cable Trays Containing Any Mixture of Cables [Cables 4/0 AWG or larger with cables smaller than 4/0 AWG] (Number of Multiconductor Cables, Rated 2000 Volts or Less, in Cable Trays)

4. What is the minimum size ladder type aluminum cable tray required for seven multiconductor cables containing 4 - 3/0 AWG conductors and four multiconductor cables containing 3 - 250 kcmil conductors? The diameter of the 250 kcmil multiconductor cable is 2.35" and the cross-sectional area of each 3/0 AWG multiconductor cable is 1.18 in^2.

Column 2 of Table 392.22(A) provides formulas for obtaining the allowed cross-sectional area of a ladder cable tray that can accommodate multiconductor cables that are smaller and larger than 4/0 AWG as well as those sized at 4/0 AWG. The formula requires the sum of the diameters (Sd) of cables that are 4/0 AWG and larger to be increased by a 1.2 inch multiplier which represents 40 percent of a standard 3 inch depth cable tray (3 in. x .40 = 1.2 in). Once this portion [1.2 **Sd**] of the formula has been determined it is then, through the process of trial and error, subtracted from a probable cross-sectional area (where the variable [**a**]* will be used to represent such as shown below) in **Column 1** until the result is larger than the total cross-sectional area of the cables that are smaller than 4/0 AWG per NEC 392.22(A)(1)(c). Again, remember that the cross-sectional areas listed in **Column 1** represent approximately 40 percent of the cable tray's allowed (usable) cross-sectional area. Also, take note that the product of the multiplier and the sum of the diameters will represent the cross-sectional area of the cable tray that's utilized by 4/0 AWG and larger cables. NEC 392.22(A)(1)(c) also requires 4/0 AWG and larger cables to be placed in a single layer and no other cables placed on them.

Again, using the formulas "as is" will require the process of trial and error until the needed result is obtained. "As is" the formula translates to,

$$\mathbf{a} - (1.2 \text{ in. x } \mathbf{Sd}) = \text{minimum cross-sectional area of required cable tray**}$$
*where **a** = probable cross-sectional area per **Column 1**
**determines required width of cable tray

so let's reconstruct the formula to produce the same results but with an easier approach. Using an easier approach, a new formula translates to

$$\mathbf{b} + (1.2 \text{ in. x } \mathbf{Sd}) = \text{minimum cross-sectional area of required cable tray,}$$
where **b** = total cross-sectional area of cables smaller than 4/0 AWG

Before using the formula, the values for **b** and **sd** must be determined.

$$\mathbf{b} = \text{(cross-sectional area of 3/0 AWG multiconductor cable)}$$
$$1.18\text{in}^2 \text{ x } 7 = 8.26 \text{ in}^2$$

$$\mathbf{Sd} = \text{(sum of diameters of 250 kcmil multiconductor cable)}$$
$$\text{Four 250 kcmil - 4 x 2.35" = 9.4"}$$

Substituting values into new formula,

$$8.26 \text{ in}^2 + (1.2 \text{ in. x } 9.4 \text{ in.}) = 19.54 \text{ in}^2$$

The results are used to determine the minimum cross-sectional area of the required cable tray per **Column 1** of Table 392.22(A). Based on the results, the maximum allowable fill area for multiconductor cables installed in a ladder type cable tray which immediately exceeds 19.54 in^2 is 21 in^2. This will require the cable tray to have an 18" minimum width. Now refer to Figure 392.22(A)(1)(c)-4.

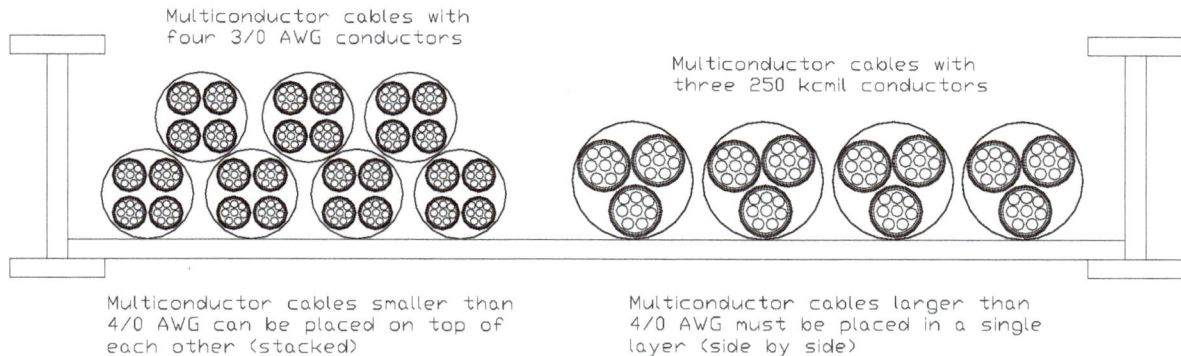

Figure 392.22(A)(1)(c)-4

If the formulas given in **Column 2** had been applied, the sum of diameters (Sd) would have required being substituted into one of the formulas until a *minimum* result exceeded the cross-sectional area of the cable smaller than 4/0 AWG. With the new formula, a *minimum* result can be obtained immediately.

392.22(A)(3)(a) - Solid Bottom Cable Trays Containing Any Mixture [Cables 4/0 AWG or larger] (Number of Multiconductor Cables, Rated 2000 Volts or Less, in Cable Trays)

5. If the tray cables in question No. 1. were installed in solid bottom cable tray, determine the minimum required width of the cable tray.

The sum of the diameters of all cables is 13.51". If these cables were installed in a solid bottom cable tray the total diameter of the cables could not exceed 90 percent of the selected cable tray width. Referring to Table 392.22(A), if a 12" solid bottom cable tray was selected, it could not be used because the total diameter of the 300 kcmil cables exceeds 90 percent of a 12" cable tray, (12" x .90 = **10.8"**[less than 13.51"]) or (13.51"/.90 = **15.01"**[width of 12" cable tray exceeded]). As a result, a cable tray with a 16" minimum width is required where 16" x .90 = **14.4"** which is greater than the diameters of all cables, 13.51".

392.22(A)(3)(b) - Solid Bottom Cable Trays Containing Any Mixture [Cables smaller than 4/0 AWG] (Number of Multiconductor Cables, Rated 2000 Volts or Less, in Cable Trays)

6. If the tray cables in question No. 3. were installed in solid bottom cable tray, what size cable tray would be required?

Considering the total cross-sectional area of the cables (9.164 in^2) and the requirements of NEC 392.22(A)(3)(b), the minimum size cable tray would be based on the maximum allowable cable fill area in **Column 3** of Table 392.22(A) that exceeds 9.164 in^2. As a result, a solid bottom cable tray with a maximum allowable cable fill area of 11 in^2 is required. This area corresponds with a cable tray having a 12" width.

392.22(A)(3)(c) - Solid Bottom Cable Trays Containing Any Mixture [Cables 4/0 AWG or Larger with cables smaller than 4/0 AWG] (Number of Multiconductor Cables, Rated 2000 Volts or Less, in Cable Trays)

7. Determine the minimum size solid bottom cable tray that's required to install the multiconductor tray cables in question No. 4.

Similar to the procedure used in question No. 4., the formulas in **Column 4** of Table 392.22(A) will be used to provide an answer for question No. 7. but will also require the process of trial and error until the needed result is obtained. The cross-sectional areas listed in **Column 3** represents approximately 31 percent (30.5 percent) of the cable tray's allowed (usable) cross-sectional area based on the multiconductor cables that are smaller than 4/0 AWG. "As is" the formula in **Column 4** translates to,

$$a - (1 \text{ in. x } \mathbf{Sd}) = \text{minimum cross-sectional area of required cable tray*}$$
where **a** = probable cross-sectional area per **Column 3** and
Sd = sum of the diameters of cables that are 4/0 AWG and larger
*determines required width of cable tray

Notice that the given formulas do not include a 1.2 in. multiplier as the previous formulas but instead applies a 1 in. multiplier (which is actually rounded-up from .9 in. which represents approximately 30 percent of the standard depth of a 3 in. cable tray [3 in. x .30 = .9 in]). Although not displayed, if the use of a 1 in. multiplier was not applied how else could the formula be applied. For example, since the sum of the diameters (**Sd**) of cables that are 4/0 AWG and larger will conclude in units of inches how could **Sd** be subtracted from the preceding areas which are in units of square inches. To subtract one unit from another both must have similar units. Therefore, to convert the **Sd** units from inches to square inches the 1 in. multiplier is applied and multiplied by the sum of the diameters **Sd** (1 in. x **Sd** [in.]). The reason why the 1 in. multiplier does not appear before **Sd** in **Column 4** is because the number 1 is a multiplicative identity meaning any number multiplied by the number 1 results to the number itself (**a** x 1 = **a**, **Sd** x 1 = **Sd**, 2 x 1 = **2**,……..). As a result, **Sd** as shown in **Column 4** is actually in units of square inches (in^2). Again, using the same approach as in question No. 4., a new formula translates to,

$$b + (1 \text{ in. x } \mathbf{Sd}) = \text{minimum cross-sectional area of required cable tray,}$$
where b = total cross-sectional area of cables smaller than 4/0 AWG

Substituting the previously calculated values (where **b** is equal to 8.26 in^2 and **Sd** is equal to 9.42") into the new formula,

$$8.26 \text{ in}^2 + (1 \text{ in. x } 9.4 \text{ in.}) = 17.66 \text{ in}^2$$

The result is used to determine the minimum cross-sectional area of the required cable tray per **Column 3** of Table 392.22(A). Based on the result, the maximum allowable fill area for the multiconductor cable installed in a solid bottom cable tray which immediately exceeds 17.66 in^2 is 18.5 in^2. This will require the cable tray to have a 20" minimum width.

392.22(A)(5)(b) - Ventilated Channel Cable Trays Containing Multiconductor Cables of Any Type (Number of Multiconductor Cables, Rated 2000 Volts or Less, in Cable Trays)

8. A 3 conductor cable consisting of 400 kcmil aluminum conductors will be installed in a ventilated channel cable tray along with two other multiconductor cables. The cross-sectional area of the 3 conductor cable is 1.897 in^2 and .854 in^2 for the other cables. Determine the required minimum cross-sectional area and corresponding width of the cable tray.

According to NEC 392.22(A)(5)(b), when more than one multiconductor cable is installed in ventilated channel cable tray the sum of the cable's cross-sectional area cannot exceed the value specified in **Column 2** of Table 392.22(A)(5). Therefore, based on the sum of all multiconductor cable,

$$1.897 \text{ in}^2 + .854 \text{ in}^2 = 2.751 \text{ in}^2$$

A ventilated channel cable tray with a maximum allowable fill area exceeding 2.751 in^2 must be used. Based on **Column 2**, the required minimum cross-sectional area of the cable tray is 3.8 in^2 which corresponds to the cable tray having a 6" width.

392.22(B)(1)(a) - Ladder or Ventilated Trough Cable Trays [cables 1000 kcmil or larger] (Number of Single-Conductor Cables, Rated 2000 Volts or Less, in Cable Trays)

9. Refer to Figure 392.22(B)(1)(a)-9. Determine the minimum width of a ventilated trough type cable tray and the ampacity of installed conductors when the cable tray will contain five 1000 kcmil THWN copper conductors, six 1250 kcmil THW aluminum conductors and four copper-clad aluminum 1500 kcmil RHW conductors. The diameters of each type conductors are 1.310", 1.539" and 1.852", respectively. All conductors are rated for 75°C and will be laid uncovered and grouped per equivalent circuits.

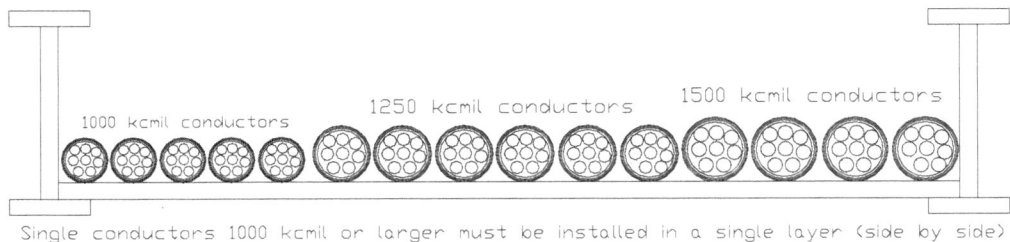

Figure 392.22(B)(1)(a)-9

Total diameter of conductors

$$
\begin{array}{l}
\text{1000 kcmil - 1.310" x 5 } = \quad 6.5500" \\
\text{1250 kcmil - 1.539" x 6 } = \quad 9.2340" \\
\text{1500 kcmil - 1.852" x 4 } = \quad \underline{7.4080"} \\
\hphantom{\text{1500 kcmil - 1.852" x 4 } = \quad} 23.1920" \text{ (total diameter)}
\end{array}
$$

WIDTH OF CABLE TRAY

Based on the total diameter, a cable tray having a width of 24" must be used in accordance with NEC 392.22(B)(1)(a) and Table 392.22(B)(1).

AMPACITY OF CONDUCTORS

The individual ampacity of each installed conductor is based on NEC 392.80(A)(2)(a) for conductors 600 kcmil and larger in uncovered cable tray. As stated, the ampacity of each conductor must not exceed 75 percent of the allowable ampacities listed in Tables 310.15(B)(17) and 310.15(B)(19). Based upon the given size of the conductors and the 75°C rating, Table 310.15(B)(17) must be referenced.

	Ampacities @ 75°C				Allowed ampacities
1000 kcmil copper	- 935A	x	.75	=	701.25A
1250 kcmil aluminum	- 855A	x	.75	=	641.25A
1500 kcmil c-c aluminum	- 950A	x	.75	=	712.50A

392.22(B)(1)(b) - Ladder or Ventilated Trough Cable Trays [cables from 250 kcmil up to 900 kcmil] (Number of Single-Conductor Cables, Rated 2000 Volts or Less, in Cable Trays)

10. Determine the minimum width of an uncovered ladder type cable tray and the ampacity of each conductor when the cable tray will contain 12-250 kcmil and 8-500 kcmil THWN copper conductors.

The cross-sectional area of each conductor is determined based on Table 5 of Chapter 9.

$$250 \text{ kcmil THWN} - .3970 \text{ in}^2 \text{ x } 12 = 4.7640 \text{ in}^2$$
$$500 \text{ kcmil THWN} - .7073 \text{ in}^2 \text{ x } 8 = \underline{5.6584 \text{ in}^2}$$
$$10.4224 \text{ in}^2 \text{ (total)}$$

WIDTH OF CABLE TRAY

Because the width of the ladder type cable tray is required to be determined according to **Column 1** of Table 392.22(B)(1), the maximum allowable fill area listed in **Column 1** which immediately exceeds the conductor's total cross-sectional area (10.4224 in^2) is 13 in^2. This will require the cable tray to be 12" in width.

AMPACITY OF CONDUCTORS

According to NEC 392.80(A)(2)(b), the allowable ampacities of single conductors in an uncovered cable tray must not exceed 65 percent for conductor's 1/0 AWG through 500 kcmil conductors. Because of the conductor's insulation type (THWN) and size, Table 310.15(B)(17) is used instead of Table 310.15(B)(19).

	Ampacity per T310.15(B)(17)				Allowed ampacities
250 kcmil copper -	405A	x	.65	=	263.25A
500 kcmil copper -	620A	x	.65	=	403.00A

392.22(B)(1)(c) - Ladder or Ventilated Trough Cable Trays [1000 kcmil and larger single-conductor cables with single-conductor cables smaller than 1000 kcmil] (Number of Single-Conductor Cables, Rated 2000 Volts or Less, in Cable Trays)

11. Five 300 kcmil THW, seven 400 kcmil THWN, four 700 kcmil RHW-2, three 900 kcmil THHN, six 1250 kcmil THW and five 1750 kcmil XHHW-2 conductors will be installed in a ladder type cable tray with a solid unventilated cover that will extend for at least 15' of the cable tray's run.

All conductors are copper with the exception of the 400 and 700 kcmil conductors which are aluminum. Determine the minimum width of the cable tray and the ampacity of each conductor type.

WIDTH OF CABLE TRAY

The requirements according to NEC 392.22(B)(1)(c) for determining the width of a ladder type cable tray *for single conductors* is similar to the requirements of NEC 392.22(A)(1)(c) for determining the width of the same type cable tray *for multiconductor cables*. **Column 2** of Table 392.22(B)(1) also provides formulas for obtaining the allowed cross-sectional area of a ladder cable tray that can accommodate single-conductor cables that are smaller and larger than 1000 kcmil as well as those sized at 1000 kcmil.

Again, the new formula previously derived in question No. 4. will be used with the exception of substituting the 1.2 multiplier with 1.1. By applying a multiplier of 1.1, the formula requires the sum of the diameters (**Sd**) of single-conductor cables that are 1000 kcmil and larger to be increased by 1.1 inches which represents approximately 37 percent (36.67 percent) of a standard 3 inch depth cable tray (3 in. x .37 = 1.1 in).

$$c + (1.1 \text{ in. x } Sd) = \text{minimum cross-sectional area of required cable tray,}$$
where **c** = total cross-sectional area of single-conductor cables smaller than 1000 kcmil

Now to determine the values for **c** and **Sd**, Table 5 of Chapter 9 is referenced to provide the cross-sectional areas and diameters of conductors.

Cross-sectional areas

300 kcmil THW	-	.5281 in^2	x	5 =	2.6405 in^2
400 kcmil THWN	-	.5863 in^2	x	7 =	4.1041 in^2
700 kcmil RHW-2	-	1.3561 in^2	x	4 =	5.4244 in^2
900 kcmil THHN	-	1.2311 in^2	x	3 =	3.6933 in^2
				c =	15.8623 in^2

Diameters

1250 kcmil THW	-	1.539 in	x	6 =	9.234 in
1750 kcmil XHHW-2	-	1.716 in	x	5 =	8.580 in
				Sd =	17.814 in

Substituting values into formula,

$$15.8623 \text{ in}^2 + (1.1 \text{ in. x } 17.814 \text{ in.}) = 35.4577 \text{ in}^2$$

The results (35.7877 in^2) will require the cable tray in **Column 1** of Table 392.22(B)(1) to have a maximum allowable fill area of 39 in^2 which reference the use of a 36" (width) cable tray.

AMPACITY OF CONDUCTORS

According to NEC 310.15(A)(3) when conductor's temperature ratings are associated together the ampacities of all conductors must be limited to the lowest temperature rating per applied table. Because the conductors involved in this installation share both 75°C and 90°C temperature ratings, the ampacity of the 90°C* conductors must be based on 75°C. NEC 310.15(A)(3) prevents conductors from being used in any way that would cause the conductor's insulation to operate under a temperature that is greater than the temperature for which the insulation was designed. NEC 310.15(A)(2) in this situation is not applicable.

The ampacities of the conductors must be determined per NEC 392.80(A)(2)(a) and 392.80(A)(2)(b).

NEC 392.80(A)(2)(a)

The ampacity of single conductors 600 kcmil and larger in cable tray with solid unventilated covers exceeding 6' shall not exceed 70 percent of its allowable ampacity. Referring to Table 310.15(B)(17) for obvious reasons, each applicable ampacity is listed and adjusted for proper use.

		Allowed Ampacity
*700 kcmil RHW-2 aluminum - 595A x .70	=	416.5A
*900 kcmil THHN copper - 870A x .70	=	609A
1250 kcmil THW copper - 1065A x .70	=	745.5A
*1750 kcmil XHHW-2 copper - 1280A x .70	=	896A

NEC 392.80(A)(2)(b)

The ampacity of single conductors sized at 1/0 AWG through 500 kcmil in cable tray with solid unventilated covers exceeding 6' shall not exceed 60 percent of its allowable ampacity. Again, referring to Table 310.15(B)(17) each applicable ampacity is listed and adjusted for proper use.

	Allowed Ampacity
300 kcmil THW copper - 445A x .60 =	267A
400 kcmil THWN aluminum - 425A x .60 =	255A

392.22(B)(1)(d) - Ladder or Ventilated Trough Cable Trays [cables 1/0 through 4/0 AWG] (Number of Single-Conductor Cables, Rated 2000 Volts or Less, in Cable Trays)

12. Five 1/0 AWG PFA copper conductors will be placed in a single layer of an uncovered ladder type cable tray. The conductors will be spaced apart at intervals no less than the diameter of one conductor. Determine the minimum required width of the cable tray and the conductor's allowed ampacity.

WIDTH OF CABLE TRAY

Referring to Table 5 of Chapter 9, the diameter of a 1/0 AWG PFA conductor is .462 in. Now the width of the cable tray will be based on the total diameter of all conductors (2.31 in.),

$$.462 \text{ in. } x \quad 5 \quad = 2.31 \text{in.}$$

and the 4 spaces between each conductor based on the diameter of one conductor ,

$$.462 \text{ in. } \times \ 4 \ = 1.848 \text{ in.}$$

Adding both results together (2.31" + 1.848") yields a sum of 4.158". The width of the cable tray must be 6" per Table 392.22(B)(1) which exceeds 4.158".

AMPACITY OF CONDUCTORS

Based on NEC 392.80(A)(2)(c) and Table 310.15(B)(19) the ampacity of the 1/0 AWG PFA copper conductors is 399A.

392.80(B)(2)(c) - Single Conductor cables (2001 volts or Over) [Triangular or Square configuration] (Ampacity of Type MV and Type MC Cables (2001 Volts or Over) in Cable Trays)

13. In sets of four, four 2.4kV single conductors are grouped together and installed in an uncovered ladder type cable tray as shown in Figure 392.80(B)(2)(c)-13. If the conductors are 250 kcmil copper, rated for 105°C, approximately 1.138" each in diameter and maintain the free air space as mentioned in NEC 392.80(B)(2)(c), determine the required width of the cable tray and the ampacity of each conductor.

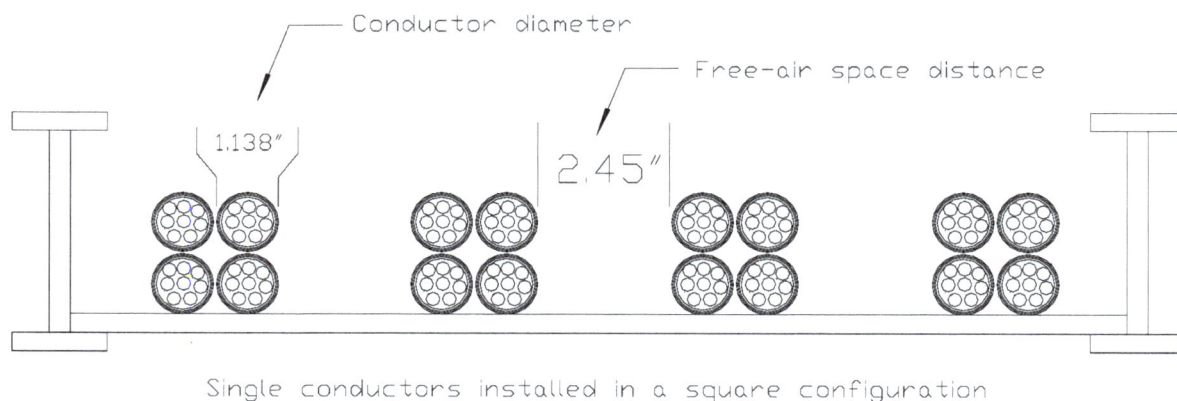

Figure 392.80(B)(2)(c)-13

WIDTH OF CABLE TRAY

As referenced in NEC 392.22(C) per NEC 392.80(B), in accordance with NEC 392.80(B)(2)(c) the width of the cable tray must be determined by the number of spaces that will exist between each set of grouped conductors. The number of spaces is then increased 2.15 times the diameter of the largest conductor contained within the configuration.

In this case the largest conductor is 250 kcmil and the configuration is a quadruplex, meaning 4 conductors per group. The requirements for considering the width of the cable tray conclude by adding the adjacent conductor's configuration to the previous. The adjacent conductors are the conductors that will actually make physical contact with the supporting structure of the cable tray. As a result, the width of the cable tray can be obtained using the following formula.

$$\text{Width} = \underline{3} \times (2.15 \times \underline{1.138"}) + \underline{8} \times \underline{1.138}$$

No. of spaces spaces between quadrupled conductors	diameter of largest conductor	No. of conductors making physical contact within the configuration	dia. of largest adjacent conductors

$$= \underline{3} \times (2.15 \times \underline{1.138"}) + \underline{8} \times \underline{1.138"} = 16.4441"$$

The required width of the cable tray must be at least 16.44".

AMPACITY OF CONDUCTORS

According to NEC 392.80(B)(2)(c) the ampacity of the conductors is determined per Tables 310.60(C)(67) and 310.60(C)(68), but only Table 310.60(C)(67) is considered because the conductors are copper. Per Table 310.60(C)(67), the ampacity of a 250 kcmil conductor at 2.4kV that's rated for 105°C is 415 amps.

Table 392.60(B) - Steel or Aluminum Cable Tray Systems (Grounding)

14. An aluminum cable tray meeting the requirements of NEC 392.60(B) is being used to distribute a variety of branch and feeder circuits throughout a processing plant. The circuits are both single and three-phases and range in ratings from 60 - 300 amps. Determine the minimum cross-sectional area of the metal (aluminum) requirements for the cable tray, in order to be used as an equipment grounding conductor.

Referring to Table 392.60(A), because the largest circuit rating of all circuits installed in the cable tray is 300 amps (and not listed), the minimum cross-sectional area of the required metal (aluminum) cable tray must be based on a 400 amps device which renders .40 in^2 for an aluminum cable tray.

ABOUT THE AUTHOR

For over thirteen years, Alvin Walker, a native of Shreveport, Louisiana owned and operated a small yet successful electrical contracting business. He now works as an author and instructor specializing in electrical and NEC training where his services are available throughout the United States. In his over thirty year of experience, he has developed a very strong background in electrical engineering, electrical design, electrical maintenance and construction. He has taught Business Law for Contractors, the National Electrical Code, Electrical Theory, and other basic and advanced electrical classes at Bossier Parish Community College, Louisiana State University-Shreveport, Northwest Louisiana Technical College (Forcht Wade Correction Center), and Southern University-Shreveport to include once serving as the Department Head of Industrial Electricity at Houston Community College-Stafford, Texas.

Mr. Walker is best known for his hands-on approach and the ability to simplify and explain the most difficult electrical subject matters. He is a master electrician and holds a Louisiana state license as an electrical contractor. He has a degree in electrical engineering from the University of South of Carolina and has worked as an electrical engineer for E.I. DuPont and Westinghouse at The Savannah River Plant (Company) of Aiken, South Carolina and M.W. Kellogg of Houston, Texas.

In his daily life Mr. Walker is a devoted Christian who has a passion for serving Christ, his fellowman and teaching and spreading the Word of God. As a recipient of three honorable discharges, he served over 9 years in the United States Army.

He enjoys traveling, wood-works and carpentry but is best known for his famous smoked barbeque ribs and sweet ice tea.

www.ingramcontent.com/pod-product-compliance
Lightning Source LLC
Chambersburg PA
CBHW042031220326
41598CB00073BA/7450